INTELLECTUAL PROPERTY IN THE 21ST CENTURY

PATENT INFRINGEMENT LITIGATION

TRENDS AND THE ROLE OF PATENT EXAMINATIONS

INTELLECTUAL PROPERTY IN THE 21ST CENTURY

Additional books in this series can be found on Nova's website under the Series tab.

Additional e-books in this series can be found on Nova's website under the e-book tab.

INTELLECTUAL PROPERTY IN THE 21ST CENTURY

PATENT INFRINGEMENT LITIGATION

TRENDS AND THE ROLE OF PATENT EXAMINATIONS

LINDSEY GONZALES

EDITOR

nova publishers

New York

Library of Congress Cataloging-in-Publication Data

ISBN: 978-1-63483-527-5

Published by Nova Science Publishers, Inc. † New York

CONTENTS

PREFACE

In August 2013, the U.S. Government Accountability Office (GAO) issued a report, for which they performed a required study on patent litigation. The study aimed to find out what was known about the extent and characteristics of patent litigation; to survey knowledgeable stakeholders about factors influencing patent litigation; to identify judicial developments that could impact patent litigation; and to examine actions taken by the United States Patent and Trademark Office (USPTO or Office) that could affect future patent litigation. The report concluded with a recommendation that the Secretary of Commerce direct the Director of PTO to consider examining trends in patent infringement litigation, including the types of patents and issues in dispute, and to consider linking this information to internal data on patent examination to improve the quality of issued patents and the patent examination process. The USPTO agreed that it would be appropriate to undertake an investigation of trends in patent infringement litigation, and to consider how any trends discovered could potentially be linked to its own internal patent examination data. The USPTO further agreed that it would be appropriate to consider whether the results of the investigation could be used to support its ongoing efforts to improve the patent examination process, and ultimately the quality of issued patents. In addition, the USPTO chose to include inter partes review (IPR) proceedings in the study, with the goal of gaining additional insights that could lead to enhanced patent quality. The USPTO carried out the investigation as recommended by GAO, and this book details the methodology and results.

In: Patent Infringement Litigation
Editor: Lindsey Gonzales
ISBN: 978-1-63483-527-5

Chapter 1

PATENT LITIGATION AND USPTO TRIALS: IMPLICATIONS FOR PATENT EXAMINATION QUALITY*

U.S. Patent and Trademark Office

EXECUTIVE SUMMARY

Introduction

In response to a recommendation from the Government Accountability Office's (GAO) Report GAO-13-465, the United States Patent and Trademark Office has conducted a study regarding the relationship between certain patent- and patent examination-related characteristics and the likelihood of subsequent patent infringement litigation initiated by the patent holder or inter partes review (IPR) petitions filed by a third party at the Patent Trial and Appeal Board (PTAB). The goal was to gather information that could lead to enhanced patent quality.

A review of the existing literature revealed that significant empirical work has been conducted relating the likelihood of litigation to various characteristics of the patents and parties involved. However, we have found few attempts to relate characteristics of the patent examination process to subsequent litigation. Also, because the IPR option is relatively new, no work

---

* This is an edited, reformatted and augmented version of a report issued January 2015.

has been done to find any systematic differences between patents that undergo IPR and those that do not. The purpose of this report is to begin to fill in some of these gaps in the existing literature, and to determine the extent to which such studies can inform improvements to patent quality.

Several factors limit the ability for us to apply the results broadly. Throughout the studies conducted in this report, we try to be careful about describing their particular limitations and about drawing proper statistical inferences.

Methods

We conducted three different types of analyses during the course of this study. First, we conducted matched case-control studies to understand the relationships between various patent- and examination-related characteristics and the likelihood of (1) subsequent patent litigation and (2) the filing of a petition for IPR. Second, we considered those patents that had been petitioned for IPR and as to which the PTAB had made a decision regarding the institution of an IPR proceeding. Using this subsample of the IPR-petitioned patents, we investigated how the likelihood of institution is related to patent- and examination-related characteristics. Third, we conducted a hands-on review of the prosecution history of all patents for which the PTAB had issued a final written decision in an IPR as of mid-July 2014. The review considered, among other things, the adequacy of the examiner's prior art search, the appropriateness of the evaluation of the prior art references, and the drafting of proper rejections.

Results

The most compelling result from the matched case-control studies of subsequent litigation and IPR petitions is that the characteristics that had the strongest relationship with each of these outcomes were those which are not directly related to the examination process itself. These characteristics include entity size, foreign origin, and the number of domestic parent applications. While the most impactful variables are defined prior to any examination, they can be indirectly affected by other policies. For instance, lower small entity fees provide a greater incentive for small entities to seek patent protection. Indirectly, this affects the rates of litigation that we later see. As another example, the examination history of a parent application could impact the characteristics of the child application that is ultimately involved in litigation.

As far as other patent- and examination-related characteristics are concerned, the clearest results are those concerning independent claims. Patents with more independent claims are more likely to be involved in subsequent litigation and IPRs. At the same time, patents with fewer words per independent claim are more likely to be litigated in court or challenged in an IPR proceeding. Holding other factors constant, an independent claim with fewer words is likely to be broader than one with more words. Thus it appears that patents with greater scope are more likely to be litigated and more likely to be petitioned for IPR. The number of independent claims may also be correlated with scope (Lanjouw and Schankerman, 2001).

Other examination-related characteristics had less explanatory power than entity size, foreign origin, and the number of domestic parent applications. Those that emerged in the matched case-control studies with some statistical significance include the General Schedule (GS)-level of the examiner, the number of information disclosure statement filings, the number of examiner interviews, and whether the application was allowed on first action (without receiving a rejection). Patents allowed by lower GS-level examiners are generally less likely to be involved in subsequent infringement litigation or IPR proceedings, as are patents allowed on first action. Both the number of information disclosure statements filed and the number of examiner interviews are positively associated with litigation, while the number of information disclosure statement filings is positively associated with IPRs. It is likely that these results are due to a selection process where applicants file more information disclosure statements and participate in more interviews in an attempt to expedite the examination process for inventions that they perceive to be more valuable. In other words, these two variables may be correlated with the value of the patent rights involved. On the other hand, applications that are allowed on the first action are likely to be ones that contain less ambiguity and uncertainty. They are likely to contain more clearly written claims. These factors are likely to reduce the amount of disputes. We hope that further research can delve more deeply into these matters. Again, it is important to note that the magnitude of the impact for the examination variables is much less than that of non-examination variables such as small entity status and foreign priority. Further, it is interesting to note that Requests for Continued Examination (RCEs), which occur during the examination process of a single application, do not have a discernable impact on litigation.

In our comparison between instituted and non-instituted IPR-petitioned patents, there were very few statistically significant results. This indicates that the observable characteristics that we identified do not do much to explain the difference between instituted and non-instituted patents. The implication is that the unobserved differences in the individual patents explain more of the variation between institution and non-institution than the observed and measured differences.

The most significant variables were related to (1) the GS-level of the examiner at the time the case was allowed and (2) the year that the patent was issued. However, even with regard to the GS-level, there is some ambiguity. We found that if the examiner who allowed the claims was a GS-13, the patent was less likely have had the IPR proceeding instituted than if the examiner was a GS-14, *or a GS-12 or below*. As far as year of issue is concerned, we found that patents issued between 2000 and 2004 were more likely to have an IPR proceeding instituted than patents issued since then.

The most notable results from the hands-on study concerned the sources of prior art used by the PTAB to find a claim unpatentable, and the examiner's prior art search. For the patents where at least one claim was held unpatentable, the prior art used by PTAB had been in front of the examiner during prosecution 59 percent of the time. Our data, while not conclusive, suggest that the number of references cited on the IDS may have been a factor. When PTAB determined that a claim was unpatentable in view of prior art not located by the examiner's search, Office of Patent Quality Assurance (OPQA) reviewers considered the examiner's search to have been deficient in 61 percent of the cases. The percentage found to have deficient searches applies only to those patents that were subject to a significant amount of filtering: they were petitioned for an IPR based on prior art not found by the examiner, the proceeding was instituted by PTAB, and at least one claim was found to be unpatentable in view of the newly submitted prior art. The self-selection of these patents, as well as possible hindsight bias on the part of OPQA reviewers, make it inappropriate to apply this figure to the average patent examination. The review also found that when examiners chose to rely on a reference as the basis of a prior art rejection, they most often do so properly. Finally, differences in claim interpretation between the examiner and the PTAB do not appear to have been a significant driver of a finding of unpatentability by PTAB.

All of the findings in this report should be interpreted in context. First, our analysis is limited by the self-selection of cases. Self-selection occurs at multiple levels for both patent infringement litigation and IPRs. First, there must be a dispute. Second, the dispute must not be easy to settle prior to seeking formal dispute resolution. All the filters of self-selection cause disputed patents to be different from the average patent. Further, only a portion of filed cases go all the way through trial, which impacts any study of trial decisions.

Finally, and importantly, changes in substantive patent law affect how we should interpret patent litigation or IPR decisions. Accordingly, if a patent was issued under examination standards mandated by then-current law, e.g., on obviousness, but litigated after changes in the law, we should be cautious about how we evaluate the examination record.

1. INTRODUCTION

In August 2013, the United States Government Accountability Office (GAO) issued GAO-13- 465 entitled "Intellectual Property: Assessing Factors that Affect Patent Infringement Litigation Could Help Improve Patent Quality" (Report). The Report was responsive to a congressional mandate in Section 34 of the Leahy-Smith America Invents Act of 2011 (AIA) that required GAO to perform a study of patent litigation. The GAO's study aimed to find out what was known about the extent and characteristics of patent litigation; to survey knowledgeable stakeholders about factors influencing patent litigation; to identify judicial developments that could impact patent litigation; and to examine actions taken by the United States Patent and Trademark Office (USPTO or Office) that could affect future patent litigation.

The Report concluded with a recommendation from GAO:

> We are recommending that the Secretary of Commerce direct the Director of PTO to consider examining trends in patent infringement litigation, including the types of patents and issues in dispute, and to consider linking this information to internal data on patent examination to improve the quality of issued patents and the patent examination process.

The USPTO agreed that it would be appropriate to undertake an investigation of trends in patent infringement litigation, and to consider how any trends discovered could potentially be linked to its own internal patent examination data. The USPTO further agreed that it would be appropriate to consider whether the results of the investigation could be used to support its ongoing efforts to improve the patent examination process, and ultimately the quality of issued patents. In addition, the USPTO chose to include inter partes review (IPR)[†] proceedings in the study, with the goal of gaining additional insights that could lead to enhanced patent quality. Even though IPR proceedings are handled by the Patent Trial and Appeal Board (PTAB) rather than a federal district court, they are adjudicative post-grant proceedings that may, in some instances, obviate the need for patent infringement litigation. Thus, IPR proceedings are appropriate for this, because they may provide additional information relevant to the goal of enhancing patent quality. Additionally, IPR proceedings are initiated by third parties, so they provide an interesting contrast to patent infringement lawsuits which are initiated by the patent holder.

The USPTO has carried out the investigation as recommended by GAO, and this response details our methodology and results. Before detailing the methodology and results, we should mention a few caveats. First, any analysis of litigated or disputed patents must recognize that those patents are not representative. The average patent is never involved in a dispute, much less litigation. Because litigated and IPR-petitioned patents are unusual, any results involving those patents should be interpreted carefully, in light of the significant self-selection of disputed patents.

Additionally, our inferences must be viewed in light of the changes that have occurred in patent law and practice over the last several years. Over the period of study, the United States Supreme Court weighed in on a number of patent cases that impact how a determination of patentability is made. In response, the USPTO issued guidance documents to ensure consistent practice by examiners, and to aid the public in understanding how the Office viewed the state of the law. In the same time period, several laws were passed by the United States Congress, including the American Inventor's Protection Act of 1999 (AIPA), the Cooperative Research and Technology Enhancement Act of 2004 (CREATE Act), and the Leahy-Smith America Invents Act of 2011 (AIA). It is important to keep these developments in mind when interpreting the results of our studies, because the laws and procedures under which patents were issued in earlier regimes may differ substantially from current practice.

The following sections detail the extent of our analysis. Section 2 provides an overview of the relevant literature and the patent examination process. In section 3, we describe the methods and data we used in two matched case-

[†] Inter Partes Review as defined by the U.S. Patent and Trademark Office is is a trial proceeding conducted at the Patent Trial and Appeal Board to review the patentability of one or more claims in a patent. See http://www.uspto.gov/patents-application-process/appealing-patent-decisions/trials/inter-partes-review for a more detailed description of inter partes review.

control studies to investigate how characteristics of the patent examination process might correlate with (1) asserting a patent in an infringement suit in federal district court, and (2) filing an IPR petition for review of the patentability of claims by the PTAB. Section 4 reports on the results of the matched case-control studies. Section 5 discusses our comparison between IPR-petitioned patents for which the proceeding was instituted and those for which it was not. In section 6, we explain our hands-on review of the prosecution history of patents which had undergone IPR. Section 7 is a brief concluding section. For all of these studies, the goal is to attempt to uncover any trends which might lead to an improved examination process and higher quality patents. This first look should be considered preliminary, and a start to what we hope will be a fruitful area of future study.

2. BACKGROUND

In this section, we provide a description of the determinants of litigation as identified by legal and economic scholars and a brief review of the relevant economic and legal literature. Because we hope to add to this literature by including measures related to patent examination, we also provide a description of some of the major milestones in the examination process. Finally, we provide a general overview of the patent examination process.

2.1. Determinants of Litigation

Theoretical models have identified several key determinants of patent litigation.[1] First, the probability of litigation increases with the probability that a patent is infringed at all. Because the infringement has to be such that it can be observed or detected by the patent owner, we would expect patents with broader claim scope to be more likely to be litigated.

Second, the likelihood of litigation increases if there is a greater divergence in parties' expectations about the outcome of a trial. This divergence of expectations can be exacerbated by uncertainty regarding the scope of patent rights and the proper boundaries around the claims being allegedly infringed. The fuzzier these boundaries are, the more likely it is that potential infringers will dispute the patent or continue their potentially infringing activities when confronted by the patent owner. Such asymmetry between the parties' beliefs may also be reflected in widely different views about appropriate licensing fees.

Third, the probability of litigation increases with the value of the patent. Litigation is costly, so patent owners are less likely to file suit if the patent is of low value. At the same time, from a theoretical point of view, low value patents are less likely to be infringed in the first place, at any given level of vigilance on the part of the patent owner. In other words, competitors are less likely to want to adopt a product or process innovation that has little or no value.

Fourth, the probability of litigation is related to the relative cost of going to trial versus the cost of settling. For certain patent owners, the cost of going to trial may be overly prohibitive. The results of Lanjouw and Schankerman (2001) suggest that this might especially be the case for foreign owners.

2.2. A Brief Literature Review

In this section we consider several studies that have examined the relationships between patent, assignee, and environmental characteristics on the one hand and the likelihood that a patent will be litigated on the other hand. In their seminal work on this subject, Lanjouw and Schankerman (2001) considered a sample of 5,452 patent cases filed between 1975 and 1991 and involving 3,887 U.S. patents. From the population of all U.S. patents (both litigated and non-litigated), they generated a control group matching on the month of the patent application and the International Patent Classification (IPC) subclass assignment. Among their findings were that roughly ten infringement or validity suits are generated for every 1,000 patents applications, domestic patent holders are more likely to file suit than their foreign counterparts, litigation rates differ by technology area, and individually-owned patents are more likely to be litigated than corporate-owned ones. They also found that litigated patents have more forward citations and forward citations per claim, suggesting that more valuable patents are more likely to be asserted in litigation cases.[2] Finally they found that litigated patents generally have more claims than their non-litigated counterparts, which they suggest indicates a link between patent scope and litigation rates.

Cockburn et al. (2003) look at "front page" information for 182 patents for which the U.S. Court of Appeals for the Federal Circuit (CAFC) had issued a ruling on validity between 1997 and 2000. The authors found that although the characteristics of patent examiners differ substantially as to experience, technological specialization, and length of time spent working on each patent application, there was no strong correlation between these characteristics and the likelihood that a patent would be invalidated by the CAFC. They noted as a "core finding" that examiners whose patents are cited more frequently tend to have a higher probability of a CAFC invalidity ruling. Cockburn et al. (2003) also concluded that although validity as determined by the CAFC is not related to the number of forward citations for the patent, validity is related to the proportion of citations attributable to an examiner's propensity to issue patents that receive a high level of citations. The examiner-specific citation rate could reflect a number of aspects of the patent examination process including differences across technologies, but the authors posited that a high degree of self-citation (the examiner's citation to patents for which she was the examiner) might be reflective of an examiner's reluctance to search beyond a narrow set of prior art with which she is already familiar.

Lanjouw and Schankerman (2004) extended their 2001 analysis by considering the effect of patent portfolio holdings on litigation activity. They obtained their data from the LitAlert database and considered 13,625 patent suits filed between 1978 and 1999. They focused on the main patent listed in each suit. A total of 9,345 such patents were included and the information included progress or resolution of suits as of the end of 1997. The method for constructing the control group was similar to the one used in Lanjouw and Schankerman (2001). The main findings of the study were that having a larger portfolio of patents reduces the probability of filing a suit on any individual patent in the portfolio and that this portfolio effect is stronger for smaller companies (measuring size by employment). For small firms, having a portfolio of patents is likely to be the key mechanism for avoiding litigation. They also found that firms operating in more concentrated technology areas (that is, where patenting is dominated by fewer companies) are much less likely to be involved in patent infringement suits. These firms are more likely to encounter the same disputants over time, so theory predicts greater incentives for settlement. Finally, they found that all sorting with respect to observed characteristics among patent disputes occurs in the decision to file suit. The key post-lawsuit outcomes do not depend on these characteristics.

In work that is most closely related to our study of the incidence of litigation, King (2003) examined the relationship between examiner hours per disposal within a particular examination group and the rate of litigation for those patents issued by the examination group.[3] His unit of analysis was the examination group itself and he focused on patents issued by each examination group in the years 1989, 1990, and 1991. He found that time spent examining was negatively related with patent litigation. In particular, his results indicated that a 1-percent increase in examiner hours per disposal is associated with a decrease in patent litigation ranging from 1.15 to 1.33 percent.

Lerner (2008) examined the litigation of all financial patents issued between 1976 and 2003.[4] He also found that there is great variation in litigation rates across technology areas as he determined that financial patents have been litigated at a rate of 27 to 39 times the rate of patents as a whole. Like Lanjouw and Schankerman, Lerner found that patents issued to individuals are much more likely to be litigated and that they appear to be more important than other financial patents in that they have more claims and more forward and backward citations. He also found that while the plaintiffs were disproportionately individual owners, defendants are usually larger firms.

In keeping with the findings of Lanjouw and Schankerman (2001), Allison (2009) also found that the characteristics that distinguish the most-litigated patents from other patents are also the ones that researchers have long used to identify the most-valuable patents: more claims, more prior art citations, more forward citations, a higher likelihood of assignment between issue and litigation, and larger numbers of continuation applications. In addition, the most-litigated patents were more likely to be software and telecommunications patents, and were disproportionately owned by non-practicing entities. These highly litigated patents also had the most continuations, and greater than 50 percent more claims than the control set.

Chien (2011) studied what she termed "acquired" or post-issuance characteristics of a patent, including changes in ownership, continued investment by way of reexamination and maintenance fees, collateralization, and forward citations, as predictors of litigation. She found that litigated patents were more likely to have changed ownership, and especially likely to have undergone a change in ownership size. They were more likely to have been reexamined, and more likely to have had their maintenance fees paid. They were also collateralized and cited more often than unlitigated patents. Petherbridge (2012) and Kesan (2012) generally confirmed the findings of Chien (2011).

Miller (2013) argued that high quality patents ex ante possess stronger property rights and are thus more likely to be found valid and infringed. In his view, patents asserted in more lawsuits should have greater litigation success because: (1) repeat patent plaintiffs choose to incur more litigation expenses and so should expect a higher return from litigation; (2) repeat patent plaintiffs tend to assert higher quality patents; and (3) divergent owner and alleged infringer beliefs about patent quality should favor the repeat patent plaintiff.

Miller (2013) found that patents asserted in more cases generally do win more validity and infringement decisions, suggesting that the higher litigation costs borne by repeat patent plaintiffs are at least somewhat compensated by the fact they assert higher quality patents.

In an area in which very little scholarly work is available to date, Love (2014) examined IPR proceedings and reported on their impact on copending litigation. Based on 979 IPR petitions filed between the September 16, 2012 inception of the procedure and March 31, 2014, Love (2014) found that IPR was an apparent improvement over the earlier inter partes reexamination procedure, which was also intended to be a quicker administrative alternative to litigation in federal court. IPR petitions were granted at a similar rate to inter partes reexamination requests. However, once instituted, an IPR was quicker than inter partes reexamination, as well as more likely to determine that the challenged claims were unpatentable. Finally, Love (2014) found that motions to stay copending litigation were more than twice as likely to be granted when the patent was involved in IPR than when it was involved in inter partes reexamination.

2.3. The Patent Examination Process

Because our study focuses on the relationship between the patent examination process and subsequent litigation, we provide a very basic description of the patent examination process and describe some of its milestones. The process begins with the filing of a patent application. When the application is received by USPTO, it goes through an extensive pre-examination review to make certain that all necessary forms have been filed, all relevant fees have been paid, and that the application is complete. A complete application requires a written description of the invention, at least one claim, and any necessary drawings.[5] As part of this review, the application is classified according to its subject matter and forwarded to the relevant technology center (TC) for examination. Within the TC, the application is then assigned to an examiner in one of the group art units (GAUs).[6] It can take several months to over a year for an application to be placed on an examiner's docket.

Examiners generally work on applications in filing date order although they have some discretion in this matter. Therefore, depending on how many applications are on the examiner's docket, an application may remain unexamined for some time even after it has reached the examiner's docket. When the examiner considers the application, she may issue a restriction requirement if multiple inventions appear in the claims. The applicant would then be required to choose claims drawn to a single invention.[7] Once an initial claim set for examination has been selected, the examiner evaluates those claims for compliance with the applicable statutes and regulations. She checks to make certain that the claims are directed to patent-eligible subject matter, that the written description is adequate to describe and enable the claimed invention, and that the claims clearly define the invention. She also conducts a search within patents and non-patent literature to find prior art references. Using these references, she determines whether the claimed invention is anticipated by a single reference, or rendered obvious either by a single reference or by a combination of references. Following this examination, the examiner may issue a Notice of Allowability that allows all claims that have been examined, or may issue a Non-Final Rejection that rejects or objects to one or more of the claims.[8] It is also possible for the examiner to issue an Office action indicating that although the subject matter of the examined claims appears to be allowable, certain formal requirements still remain and must be addressed prior to allowance. Between 85 and 90 percent of all applications receive a Non-Final Rejection in the first Office action.[9]

The applicant is generally given three months to respond to a non-final Office action, but may take up to three additional months in exchange for additional fees. The applicant typically responds with some combination of arguments and amendments to the claims to clarify them or to narrow their scope to avoid the prior art. The applicant may also file information disclosure statements (IDSs), which are used to comply with the applicant's duty to disclose any information material to patentability. The information typically includes relevant prior art, particularly art revealed to the applicant during the examination of a related foreign or domestic application. The applicant may also ask for a telephonic or in-person interview with the examiner. After the examiner receives the applicant's response, she reevaluates the claims to determine whether the rejections or objections have been overcome. If no issues remain, the applicant is informed that the claims are allowable. Otherwise, the examiner will typically issue a Final Rejection, thus formally closing the examination process—at least temporarily.

After receiving a Final Rejection, the applicant has several options. First, the applicant may choose not to continue to seek patent protection for the invention by abandoning the application, either by express request or simply by failing to respond within the specified period.[10] Second, the applicant may continue to seek patent protection before the examiner. This may be done either by filing a continuation application (CON) which is entitled to benefit of the filing date of the original application,[11] or by filing a Request for Continued Examination (RCE).[12] Finally, the applicant may file an appeal with the USPTO's PTAB[13] arguing that the PTAB should reverse the examiner's rejections.

If examination continues before the examiner, the applicant has further opportunities to amend claims and make further changes. Again, the examiner may or may not allow the claims and may issue further Non-Final Rejections and Final Rejections. The applicant can again respond, and this process can go through several rounds. On average, each round of examination tends to lead to changes in the application's claims. Also, it is important to note that there is no such thing as a terminal rejection. Prosecution terminates with either an issued patent or an abandonment. For the applications that we will be considering in our empirical section, at least some of the claims were allowed and the patent was issued.

3. Matched Case-Control Studies

Following Lanjouw and Schankerman (2001, 2004) we conducted matched case-control studies on the relationships between particular characteristics of patent examination and the likelihood of either (1) the filing of a patent infringement suit or (2) the filing of a petition for inter partes review. Matched case-control studies entail the matching of a control group to a sample from a population of interest. In our case we start with samples from the populations of patents that have been litigated or have been petitioned for IPR. This study design has generally been used as a way to reduce confounding in the sampling of the control group. However, more recently, statisticians have argued that the true advantage of match case-control studies is greater statistical efficiency.[14]

To compare the matched samples, we use conditional logistic regression analysis, which estimates the relationships between certain explanatory variables and the likelihood that a patent will be involved in infringement litigation, or be petitioned for IPR. Logistic regression is one statistical method designed to analyze categorical data. In this case, there are two categories. Our outcome variable of interest – whether or not the patent was asserted in an infringement suit (or was the subject of an IPR petition) – can only fall into one of two categories: yes or no. The logistic model is designed for situations such as this (Maddala, 1983). We describe this method more in the analysis section.

Below, we start by describing our samples of litigated and IPR-petitioned patents and the USPTO data that we joined to these samples. We then describe the methods that we used to create matched control groups of unlitigated and unpetitioned patents. We close this section by describing the explanatory variables used in the statistical analysis.

The source data used in the analyses come from several different sources. To create a list of litigated patents, we use data obtained by the GAO detailing 500 patent infringement suits filed between 2007 and 2011.[15] To create a list of IPR-petitioned patents, we use a list of all complete IPR petitions provided by the PTAB. Most of the patent examination-related variables are generated using data from USPTO's Patent Application Location and Monitoring (PALM) system. Other data providing information regarding claim construction and examiner seniority are also used. In this section, we provide more information on each of these sources and how elements from each are combined to create our analysis file.

3.1. Data Sources

3.1.1. GAO Litigated Patent Data

As part of its study of patent litigation, GAO (2013) acquired data from Lex Machina, which is a company that provides intellectual property data and analytical services to law firms and corporations.[16] These data include a random sample of 100 patent infringement filings in federal district courts for each year from 2007 through 2011, for a total of 500 filings. The filings represent roughly 4 percent of all lawsuits that can be found in the Lex Machina database covering this 5-year period. For each filing, the patent(s) being litigated were identified by Lex Machina. GAO, with Lex Machina's permission, provided the USPTO with these data for the purposes of this study. Among the 500 lawsuits, 992 unique patents were involved. It was this list of litigated patents to which we merge USPTO data on patent- and examination-related characteristics and create a control group of other patents. It is important to recognize that when we use the term "litigated patent" in this document, we mean a patent that was asserted in an infringement complaint, regardless of whether or not the case settled prior to trial. We believe that this usage comports with that of the GAO Report.

3.1.2. PTAB Inter Partes Review Data

IPR is a new trial procedure instituted under the AIA that may be used to reassess the patentability of claims in an issued patent in view of the prior art. IPR, which is handled by the PTAB rather than a federal district court, is intended to be a faster, less expensive alternative to litigation. The new proceeding went into effect on September 16, 2012, and may be used for any U.S. patent regardless of its issue date. An IPR proceeding begins with a petition brought by a third party (i.e., someone who is not the patent owner). The third party petitioner must show that there is a reasonable likelihood that at least one claim in the patent is anticipated by or obvious over one or more prior art patents or printed publications. If the petitioner's showing is adequate, the PTAB will institute the IPR proceeding.

The PTAB provided us with a list of all completed petitions filed up to mid-July 2014. The file also includes information such as the name of the petitioner, the patent owner, the patent number, the number of claims being challenged, and any PTAB rulings. The PTAB rulings include the number of claims for which the IPR proceeding was instituted. The data set consists of 1,537 petitions covering 1,040 patents. As with the list of litigated patents from the GAO, we merge USPTO data on patent- and examination-related characteristics and create a control group of other patents for which no IPR petition had been filed.

3.1.3. PALM Data

The PALM system is used by patent examiners to monitor the progress of patent application prosecution. It contains a list of all communications either sent or received by the USPTO in connection with prosecution of a particular patent application. It also includes additional information concerning the internal processing of the application. For instance, the file includes information about when each application was filed, and when it was placed on an examiner's docket.[17] PALM data indicate whether the application is for a utility, design, or plant patent, and the group art unit to which the application is assigned. The PALM database also includes the nature and date of each of the examiner's Office actions and each of the applicant's responses. Office actions include, but are not limited to, Non-Final Rejections, Final Rejections, and Notices of Allowability. The applicant's submissions can include arguments, amendments, IDSs, and Notices of Appeal. If the applicant files an RCE, the RCE filing is reflected in PALM.[18] PALM also includes information about related applications, i.e., when the application is the national stage of an international application, or when it has a domestic or foreign parent application. Information about the applicant is included, such as name, address, citizenship, representation by counsel, and whether the applicant qualifies for small entity status. Finally, information about issuance of a patent or other final disposition of an application may also be found in PALM.

3.1.4. Other USPTO Data Sources

Although PALM is our primary source of data on patent characteristics, certain metrics that we include are available only from other sources. Information on forward citations (used for matching) comes from the Technology Assessment and Forecast (TAF) database, which is administered by USPTO's Patent Technology Monitoring Team (PTMT).[19] We also look at the actual claims for each patent in our data set to generate measures such as each patent's number of independent claims and the number of words per independent claim. We obtain the claims information from the Patent Grant Red Book XML files maintained by Reed Tech.[20] Finally, we use internal USPTO human resources data on promotions to determine the pay grade of the examiner to whom the case was docketed at the date of patent allowance.

3.1.5. Database Construction

Figure 3-1 illustrates how the data from the various sources are combined to generate a file that contains information on the pertinent characteristics of the litigated patents from the GAO sample. The data are combined similarly for patents identified as having been petitioned for IPR in the PTAB file.

In general, most of the data are merged on the patent numbers of the litigated and IPR-petitioned patents. The merging for examiner pay grade data is more complex. PALM includes information on the various examiners who worked on each patent application, and the dates over which each examiner was assigned to it. From this information, one can determine which examiner was assigned a patent application on the day that it was allowed.[21] The human resources data include each examiner's GS-level and the date range over which the examiner was at the GS-level. Using this information we merge on the examiner identification number and the allowance date to determine the GS-level of the examiner at the time of allowance.

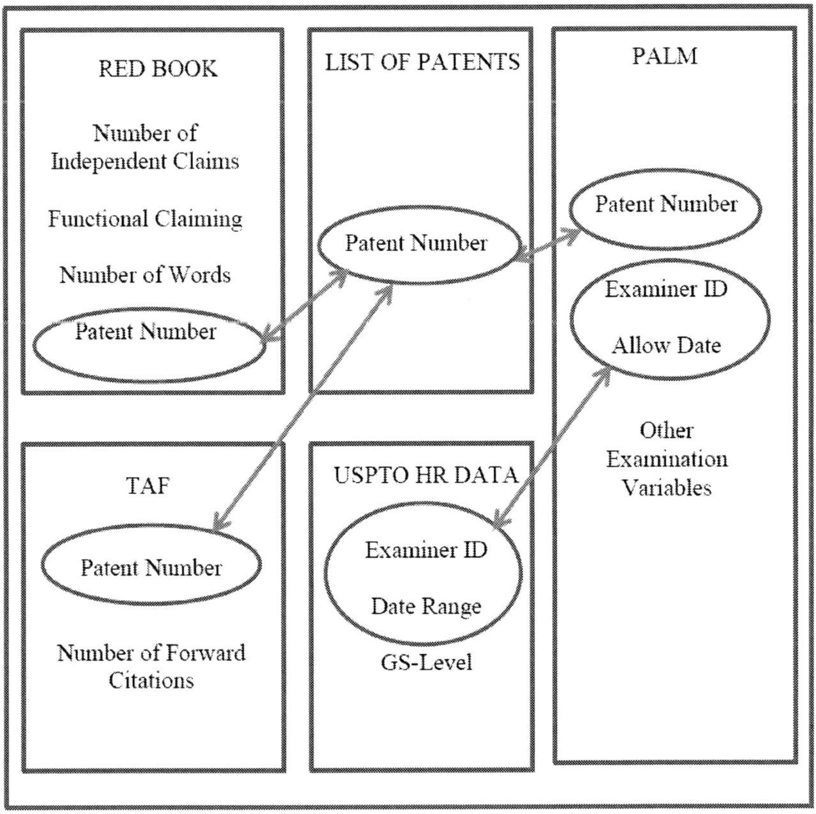

Figure 3-1. Data Set Construction.

3.2. Matched Samples

As we stated earlier, the lists of litigated and IPR-petitioned patents are provided by the GAO and the PTAB, respectively. Our ultimate goal is to determine whether these patents differ from unlitigated and unpetitioned patents in any systematic way with respect to the patent examination process at USPTO. In order to do that, we conduct a one-to-one matched case-control study. For each subject patent, we select one control patent that is similar to the subject patent, but for which we do not observe litigation or petition for IPR. The extent of the similarity depends upon the particular matching method that is used. In this analysis we use two different methods, and compare the results.

3.2.1. Matching Algorithms

Below, we describe the details for generating the matched samples with regard to the incidence of litigation. We follow the same mechanism for generating the matched files for the incidence of petition for IPR, except that we begin with the list of IPR-petitioned patents from the PTAB. It is important to note that any characteristic used for matching cannot be used in the statistical analysis. For instance, we require a matched patent to have the same maintenance payment history as its litigated partner. Thus, there will be no variance in the maintenance histories across the litigated group and the control group. Each study design must determine at the outset the set of variables that are used to determine a good match, in order to isolate the variables of interest in the study.

1. Define the set of all potential matches.

We create a comparison group of unlitigated patents that are identical to the litigated patents from the GAO sample across three dimensions. These dimensions include: (1) the year of the patent grant, (2) the workgroup to which the patent was assigned on the date of allowance, and (3) the patent's maintenance fee history.[22] This results in multiple, often thousands of, patents that could be matched to each of the individual patents in the GAO sample – the potential matches.

2. Choose one control patent for each litigated patent as defined by the particular matching algorithm.

In order to choose the closest match among the many potential matches we use two different algorithms, which lead to the generation of two different control groups.[23] Both algorithms rely on a "nearest neighbor" criterion; however, they differ in the characteristics used to determine determining the nearest neighbor.

a. Citation match.

The first algorithm makes a simple match on one patent characteristic: citations. In particular, it matches each patent from the GAO sample to that member of its potential matches that is closest to the litigated patent with respect to the number of citations received within the first three years of issue.[24] We match on the basis of citations because we would like to control for patent value as much as possible, in order to isolate the differences in patent examination. Patent maintenance payments and patent citations are two commonly used correlates of value in the economics literature. As a simplified example, consider a litigated patent – patent '101 – with 30 such forward citations. Suppose that the process described in Step 1 had generated a list of three potential matches: patents '201, '202, and '203. Further, suppose that patent '201 had received 24 forward citations, patent '202 had received 10 forward citations, and patent '203 had received 35 forward citations. In this case, the algorithm would choose patent '203 as the final unique match for patent '101, since the difference in the number of forward citations is smaller for patent '203 (5) than it would be for patent '201 (6) or patent '202 (20). In the event of ties, the matched un-litigated patent is randomly chosen from among those patents that were tied.

b. Propensity score matching.

The second algorithm uses a propensity score matching algorithm to choose from among the potential matches on the basis of patent citations as well as several other characteristics.[25] In propensity score matching, the litigated patents and their potential matches are first combined into one data set. Next, a classification model – in our case a logistic model – is estimated using a number of patent characteristics as potential predictors of litigation. These predictors include continuation history, small entity status, claims of foreign priority, claimed parent type (relationship to an earlier U.S. application), the number of all (independent and dependent) claims at application filing, the number of citations received in the first three years after issue, year of issue, interactions between the year of issue and three-year forward citations,[26] and the technology categories defined by Hall, Jaffe, and Trachtenberg (2001). Most of these characteristics are known at the time of filing. Matching on these allows us to focus on examination-related characteristics in the subsequent analysis. After the model is estimated, it is used to predict the probability that any patent in the entire data set would have been a member of the litigated set and this probability is converted into a propensity score. The last step is to match each patent from the GAO sample to that member of the set of its potential matches that is the nearest neighbor to the litigated patent with respect to the propensity score.

3.2.2. Matched Litigation Sample

Our sample consisted of 975 patents that were named in the 500 sampled patent litigation filings between 2007 and 2011. Figure 3-2 illustrates the mix of patents in the sample by year of issue and general technology area.[27] The technology area designations are defined using the TC at USPTO where the patent had been examined. They are defined as follows:

- Biotechnology and organic chemistry (BIO) – currently TC 1600[28]
- Chemical and materials engineering (CHEM) – currently TC 1700[29]

- Computers and communications (COMP) – currently TCs 2100, 2400, and 2600[30]
- Semiconductors, electrical and optical systems and components (SEMI) – currently TC 2800[31]
- Transportation, construction, electronic commerce, agriculture, national security and license & review (TRANS) – currently TC 3600[32]
- Mechanical engineering, manufacturing, products (MECH) – currently TC 3700[33]
- Designs (DES) – currently TC 2900

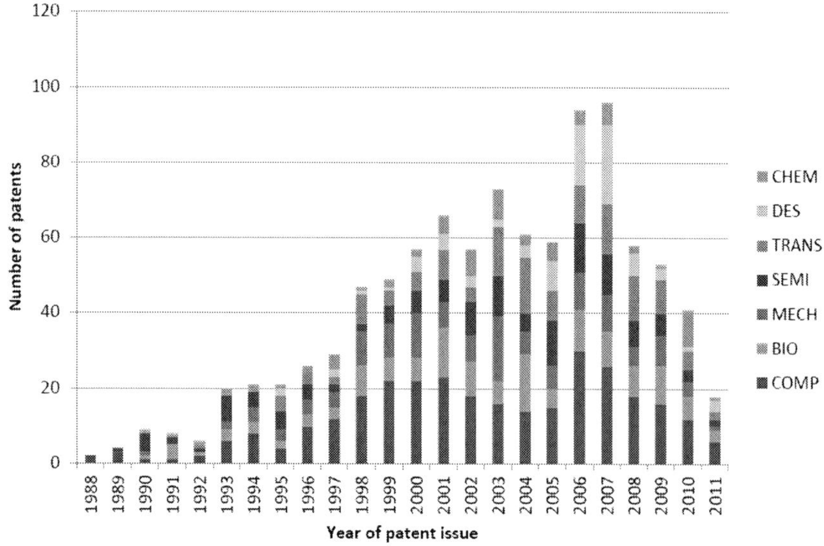

Figure 3-2. The Mix of Patents from the GAO Sample by Year of Issue and Technology Area.

Not surprisingly, the sample is weighted heavily toward recently-issued patents. Seventy five percent of the patents were issued after 1999, roughly half of the patents were issued after 2003, and slightly more than one-quarter of the patents were issued between 2007 and 2011. As far as technology area is concerned, roughly 30 percent of the patents in the sample come from the COMP area, while roughly 13 to 14 percent come from each of the following four areas: BIO, MECH, SEMI, and TRANS. DES and CHEM patents make up the remaining 15 percent of the distribution in roughly equal numbers.

The initial set of potential matches for the GAO sample was generated by doing an exact match on year of issue, the workgroup to which the patent application had been assigned for examination (which is a subset of the TC), and the maintenance history of the patent. We then generated matched control groups using nearest-neighbor matching either on the number of citations received within three years of patent issue or on propensity scores as described in the methods section above. Here we briefly describe how closely we were able to match on these two factors.

Table 3-1. Distributions of Matching Factors for the GAO Sample and the Control Group

	Number of Citations	
Percentile	Litigated	Control Group
Smallest	0	0
1%	0	0
5%	0	0
10%	1	1
25%	2	2
50%	6	6
75%	15	15
90%	30	30
95%	47	47
99%	119	117
Largest	323	335
Mean	12.9	12.9
St. Deviation	21.9	22.1

		Propensity Score	
Percentile		Litigated	Control Group
	Smallest	0.0000466	0.0000466
	1%	0.0000958	0.0000959
	5%	0.0002313	0.0002278
	10%	0.0003472	0.0003472
	25%	0.0006138	0.0006144
	50%	0.0012682	0.0012682
	75%	0.0025817	0.0025954
	90%	0.0039935	0.0039911
	95%	0.0058683	0.0057706
	99%	0.0200807	0.0187193
	Largest	0.0552789	0.0463568
Mean		0.0021186	0.0020724
St. Deviation		0.0035318	0.0031955

In Table 3-1, we present the distributions of the three-year forward citations and the propensity scores for the GAO and the two control groups. In each case, we were able to generate a control group that closely matches the GAO sample. Matching on three-year forward citations, we were able to generate an exact match for 938 of the 975 GAO patents. For 28 of the remaining 37 patents, we were able to generate a match where the difference in forward citations was two or less, which is quite good considering that all of these sampled patents had more than 50 citations. The distributions of the propensity score for the GAO sample and control group are also nearly identical. The average difference in the score between the two groups (0.00008) is quite small compared to the average score of 0.002.

3.2.3. Matched IPR Sample

Our sample of IPR-petitioned patents consists of 1,014 patents that were named in completed IPR petitions between September 16, 2012 when the procedure became available, and mid-July 2014. Figure 3-3 illustrates the mix of patents in the sample by year of issue and general technology area.[34]

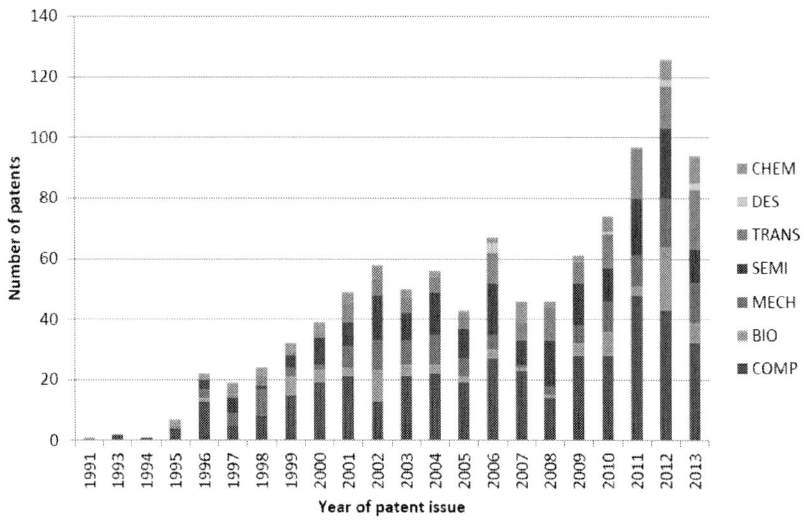

Figure 3-3. The Mix of IPR-Petitioned Patents by Year of Issue and Technology Area.

The great majority of the IPR-petitioned patents have been issued very recently. Half of the patents were issued after 2007, and slightly more than 30 percent of the patents were issued between 2011 and 2013. As far as technology area is concerned, roughly 40 percent of the patents in the sample come from the COMP area, while roughly 20 percent come from the SEMI area. The BIO and CHEM areas combine to account for 15 percent of the patents, while 25 percent of the patents are in the MECH and TRANS areas. There are only 8 DES patents in the PTAB sample, accounting for about 0.8 percent.

Table 3-2. Distributions of Matching Factors for the PTAB Sample and the Control Group

	Number of Citations	
Percentile	IPR	Control Group
Smallest	0	0
1%	0	0
5%	0	0
10%	0	0
25%	1	1
50%	5	5
75%	13	13
90%	32	31
95%	54	53
99%	105	104
Largest	310	227
Mean	12.4	12.1
St. Deviation	24.3	22.3

Percentile	Propensity Score	
	IPR	Control Group
Smallest	0.0000438	0.0000438
1%	0.0000701	0.0000732
5%	0.0001948	0.0001948
10%	0.0003290	0.0003288
25%	0.0006077	0.0006078
50%	0.0012391	0.0012305
75%	0.0023629	0.0023620
90%	0.0038905	0.0038821
95%	0.0051055	0.0049758
99%	0.0104025	0.0100328
Largest	0.5337020	0.0925166
Mean	0.0025418	0.0019779
St. Deviation	0.0175829	0.0042253

As with the GAO sample, the initial set of potential matches for the PTAB sample of IPRs was generated by doing an exact match on year of issue, the workgroup to which the patent application had been assigned for examination (which is a subset of the TC), and the maintenance fee payment history of the patent. Matched control groups were then generated using nearest-neighbor matching either on the number of citations received within three years of patent issue or on propensity scores as described in the methods section above. Here we briefly describe how closely we were able to match on these two factors. Table 3-2 presents the distributions of the three-year forward citations and the propensity scores for the PTAB and the two control groups. In each case, we were able to generate a control group that closely matches the PTAB sample of petitioned patents. Matching on three-year forward citations, we were able to generate an exact match for 971 of the 1,014 PTAB patents. For 27 of the remaining 43 patents, we were able to generate a match where the difference in forward citations was two or less, which is quite good considering that 25 of these 27 sampled patents had more than 20 citations each. The distributions of the propensity score for the PTAB sample and control group are also nearly identical, except for the extreme right tail of the distribution where there are a few petitioned patents with very large propensity scores.[35]

3.3. Explanatory Variables

After generating the matched samples we estimate a model meant to predict the likelihood that an "event" will take place using information regarding each of the patents in our matched sample. The event can be either the filing of a patent infringement lawsuit by a patent owner or the petition for an IPR by a third party. In addition to the examination-related variables of most interest, the basic model also includes indicator variables for each of the matched pairs.[36] In what follows, we describe the explanatory variables included in the statistical models. We chose these variables because they can be created for all patents without resorting to hand coding and because they represent important aspects of the patent examination process. They can be broken out into three main categories:

- Characteristics of the application at the time of filing.
- Characteristics of the examination.
- Characteristics of the allowed claims.

3.3.1. Application Characteristics

The first set of variables includes characteristics of the application, most of which are known at the time of filing. As a control, we include year of application filing indicators. This set of variables identifies applications as having been filed between 1981 and 1990, between 1991 and 1995, between 1996 and 2000, between 2001 and 2005, or after 2005.

Small Entity Status

This indicator variable is set equal to one in cases where the filer was granted small entity status and set to zero otherwise. Lanjouw and Schankerman (2001) have found that individually owned patents are more likely to be litigated than corporate owned ones, despite that fact that individuals likely face steeper litigation costs. They suggested that this would most likely be due to corporate owners' advantages in reaching settlement agreements before having to file suit. We use the small entity indicator to control for any possible differences in litigation rates.

Number of Parent Applications

In some cases an application claims the domestic benefit of an earlier-filed U.S. application. Applications earlier in the continuity chain are usually called parent applications. Thus, this variable is a count of the number of regular U.S. parent applications (i.e., non-foreign, non-provisional applications) in the chain of continuations from the progenitor application to the subject patent. Parent applications may or may not have issued as separate patents.

Pendency

This variable also relates to the continuation history of the patent. In this case we measure pendency (in years) from the application date of the earliest parent application (the progenitor application) to the docketing date of the issued application. For an application with no parents, this variable is equal to the pendency from the application date to docketing.

Foreign Origin

A final application characteristic that we consider is whether the application was filed by a foreign entity. We consider two different proxies for foreign applications. First, we include a variable indicating whether the application claims priority to a foreign application under 35 U.S.C. § 119. Second, we include an indicator for applications that are U.S. national stage filings of Patent Cooperation Treaty (PCT) applications under 35 U.S.C. §

371.[37] Previous studies, including those of Lanjouw and Schankerman (2001), found that U.S. patents granted to foreign applicants are less likely to be litigated.

3.3.2. Examination Characteristics

As far as examination history-related variables are concerned, we consider several different metrics that are related to the intensity of activity in an application.

IDS Filings

This variable is a simple count of the number of instances in which an IDS is recorded in PALM. More IDS filings may indicate greater "effort" by the applicant and thus be a signal of higher value. Higher value patents are more likely to be litigated. Note that this is a count of the number of IDS forms submitted, and not a count of how many document are cited on those IDS forms.

Number of Interviews

This variable is a simple count of the number of instances in which an interview is recorded in PALM.[38] Like IDS filings, interviews may be correlated with applicant investment in the invention and signal higher value.

Examination Pendency

We include a measure of each application's examination pendency, defined as the time (in years) between docketing and allowance. We would expect on average that an application with longer examination pendency would have gone through more rounds of negotiation between the applicant and the examiner.[39] The next three variables are all related to the idea of repeated negotiation.[40]

RCE Count

This is a simple count of the number of RCEs (or RCE-type continuation) filed during the course of examination. RCEs are most often utilized by applicants as a means for reopening the examination of a patent application that has received a Final Rejection. There is no formal limit to the number of RCEs that an applicant can file in a particular application, although there is a fee involved.[41]

Appeals

Another measure that we consider is the use of appeals to the PTAB or to its predecessor, the BPAI. In order to account for this, we include an indicator equal to one for cases where there was at least one appeal that ended with a PTAB decision. This choice was made to distinguish those cases that went all the way through the appeal process from those where the appeal is filed, but later withdrawn. In the latter case, the appeal may be used as a delay tactic.

First-Action Allowance

This is an indicator variable set equal to one for cases in which the application is allowed on the first action. In these cases, the examiner, finding no grounds for rejection or objection, allows the claims without ever having issued a rejection. This does not indicate that the claims at grant are identical to the claims at filing; however, it is unlikely that there are substantial changes. For the purposes of defining this variable, we treat continuations (CONs and CIPs) and divisional applications (DIVs) identically to new applications. That is, a CON can have a first-action allowance.

Examiner Seniority

To account for examiner characteristics, we define a set of four indicator variables related to the pay grade (at the time of allowance) of the examiner who allowed the patent. We break out the examiners into three categories. The first category consists of examiners having a pay grade of GS-12 or below. These non-signatory examiners cannot represent the USPTO or independently sign any of their own work, so a supervisor or primary examiner with full signatory authority must review the work and sign it. Given that their work is reviewed by a supervisor or primary examiner, there may be no substantive difference between patents examined by non-signatory examiners. In addition, having two examiners involved in the examination may provide an independent benefit.

The second category consists of examiners at the GS-13 level. Many of these examiners have partial signatory authority, or even temporary full signatory authority, and therefore can sign at least some of their own work. They tend to have more responsibility and are supervised less closely than examiners at lower GS-levels. These examiners are also being reviewed on the Partial and Full Signatory Authority Program by a panel of supervisors. So, their work is highly scrutinized – thus, we expect them to be less likely to allow claims that may end up in litigation.

The third category consists of full signatory examiners with pay grades at GS-14 or above. Although supervisors still provide oversight, these primary examiners are generally expected to be able to work independently.[42] As examiners progress through the GS-levels they gain experience. Therefore, as the GS-level increases, the time allotted for an examiner to complete her review of an application decreases. Information about the examiner's pay grade was missing for roughly 5 percent of the patents in our sample, so we created an indicator variable for such cases.

3.3.3. Characteristics of Allowed Claims

Finally, we include variables meant to represent the characteristics of the allowed claims for each of the patents in our data sets.

Number of Independent Claims

The first variable is a count of allowed independent claims. We expect this variable to be positively correlated with litigation. Lanjouw and Schankerman (2001, 2004) use the number of claims as one proxy for patent scope, which they find to be positively correlated with litigation.[43] One complication with this interpretation is that applicants may choose to include narrow independent claims that may even be subsets of broader independent claims. So, additional independent claims do not *necessarily* increase the scope of the patent. More typically, additional independent claims cover different aspects of the invention. For instance, an applicant may claim a method, and may separately claim a computer-readable medium that contains executable instructions to perform the method. The applicant may further claim a system configured to perform the method. For high value inventions, the applicant may want to invest in overlapping independent claims that cover slightly different sets of possible embodiments.[44]

Words Per Independent Claim

The second variable is the average number of words per independent claim. We expect that as the number of words per independent claim increases, the boundary of the property right would be better defined. Thus it is our hypothesis that this variable is negatively correlated with litigation. Better-defined boundaries would lead to fewer disagreements regarding whether a party's actions amounted to infringement.

Functional Claiming

We also include an indicator for whether any of the claims have a functional limitation. A functional claim limitation is commonly introduced by the phrase "means for" or "step for." Although it is possible that a functional claim limitation may lack one of these phrases, or even that a non-functional claim limitation may include them, for the purpose of this study we considered a claim to include a functional limitation if one or both of the phrases "means for" or "step for" appeared in it. Using functional claim language allows the claim drafter to rely on material in the specification to define the scope of protection afforded by a claim, without having to spell out all of the details explicitly in the claim itself. Therefore, claims that contain functional limitations are sometimes considered to be less clear than claims that do not.

4. Estimation of Matched Case-Control Studies

In this section we present the results regarding the relationships between patent- and examination-related variables and the likelihood of either (1) the filing of a patent infringement suit or (2) the filing of a petition for litigation or inter partes review. In each sub-section below, we provide some basic comparisons between the sampled patents and their matched control groups, and the results of the relevant econometric models.

Biostatisticians have long suggested the use of a particular variant of logistic analysis, conditional logistic analysis, when conducting matched case-control studies (Hosmer et al., 2013). This is particularly important when one is using a 1-to-1 matched design. As an example, consider this study. We are investigating the relationship between up to 21 different factors and the likelihood of litigation. Our matching algorithm generated 975 matched pairs that can be included in the analysis. So, in a fully stratified analysis we are required to estimate 996 parameters consisting of a constant term, the 21 factors of interest and 974 stratum (or matched-pair) indicators.[45]

We use conditional logistic regression analysis to estimate the relationships between the explanatory variables described above and the likelihood that a patent will be involved in infringement litigation (or be petitioned for IPR). We choose logistic regression analysis because our outcome variable of interest – whether or not the patent was litigated (or was the subject of an IPR petition) – can only fall into one of two categories: yes or no. In addition, these categories are essentially qualitative rather than quantitative in nature. The logistic model is designed for situations such as this (Maddala, 1983).

All results from the logistic regressions are reported as odds ratios. The odds that an event will occur are defined as p/(1-p), where p is equal to the probability that the event occurs. In a logistic regression, an odds ratio of 1.2 for a variable indicates that a 1-unit increase in the value of that variable would be associated with a 20-percent increase in the *odds* that the event occurs. An odds ratio of 0.8 indicates that a 1-unit increase in the value of the variable would be associated with a 20-percent decrease in the *odds* that the event occurs. Generally, an odds ratio greater than 1 indicates a positive relationship between the variable and the likelihood of the event occurring, and an odds ratio of less than 1 indicates a negative relationship between the variable and the likelihood of the event occurring.

4.1. Incidence of Litigation

4.1.1. Means Comparisons

Although our statistical analysis focuses on the use of logistic regressions, it is instructive to first examine some summary statistics. Table 4-1 compares the means of the explanatory variables for the GAO sample and those for the citation-matched control group.

Table 4-1. Means Comparisons for Litigated Patents and Three-Year Citation-Matched Control Group

	Litigated Patents (n=975)		**Control Group** (n=975)	
Examiner Seniority	**Mean**	**Std Error**	**Mean**	**Std Error**
GS-12 and Lower***	0.217	0.013	0.274	0.014
GS-13	0.161	0.012	0.177	0.012
GS-14 and Higher***	0.568	0.016	0.490	0.016
GS Level Missing	0.053	0.007	0.058	0.008
Small Entity* **	0.346	0.015	0.191	0.013
Continuation History				
Number of Domestic Parents***	1.823	0.122	0.689	0.044
Pendency from Earliest Parent to Docketing***	1.575	0.051	0.926	0.033
Foreign Applications				

	Litigated Patents		Control Group	
	(n=975)		(n=975)	
Examiner Seniority	**Mean**	**Std Error**	**Mean**	**Std Error**
Foreign Priority Claim***	0.130	0.011	0.292	0.015
371 Case***	0.031	0.006	0.061	0.008
Examination Variables				
Number of IDS Filings***	3.889	0.156	3.005	0.122
Number of Interviews***	0.419	0.027	0.294	0.022
Pendency Before Examiner	1.735	0.047	1.837	0.047
Number of RCEs	0.294	0.023	0.245	0.019
At Least One Appeal	0.021	0.005	0.016	0.004
FAOM Allowance***	0.136	0.011	0.182	0.012
Patent Claims Variables				
Number of Independent Claims***	4.117	0.162	3.156	0.089
Number of Words per Independent Claim***	137	3	150	4
Functional Claim***	0.265	0.014	0.212	0.013

*** Indicates that the mean values are significantly different at the 1-percent level.

The results of the means comparisons indicate the following:

- Litigated patents are less likely to have been allowed by examiners of GS-12 or below and more likely to have been allowed by examiners of GS-14 or above. Twenty-two percent of the litigated patents were allowed by examiners of GS-12 or below compared to 27 percent of the patents in the control group. At the same time, 57 percent of the litigated patents were allowed by examiners of GS-14 or above compared to 49 percent of the control group patents. Both of these differences are statistically significant at the 1- percent level.[46]
- Litigated patents are more likely to have been granted to small entities. Roughly 35 percent of the litigated patents were granted to small entities, while only 19 percent of the patents in the control group were granted to such entities. This result is consistent with results found by Lanjouw and Schankerman (2001).
- Litigated patents are more likely to have an extensive prior continuation history. The average number of domestic parents for litigated patents is more than two times the number for patents in the control group. Also, the average pendency from the filing date of the earliest parent to the docketing date of the later-issued application is roughly 70 percent longer for litigated patents.

- Litigated patents are less likely to have been granted on an application with a foreign or international priority. Either way that we measure foreign applications – a foreign priority claim or a 35 U.S.C. § 371 national stage entry – we get the same result. Only 13 percent of the litigated patents had evidence of a foreign priority claim compared to 29 percent of the patents in the control group. Likewise, 3 percent of the litigated patents were 35 U.S.C. § 371 national stage entries compared to 6 percent for the control group.
- Litigated patents have a higher number of IDS filings and examiner interviews during examination. The average number of IDS filings for litigated patents was 30 percent higher than the average for the control group. Likewise, the number of examiner interviews for litigated patents was, on average, more than 40 percent higher than the number of interviews for the control group.
- Litigated patents are less likely to have resulted from a first-action allowance. Fewer than 14 percent of the litigated patents had been allowed on first-action compared to roughly 18 percent of the patents in the control group.
- Litigated patents tend to have more independent claims, fewer words per independent claim, and more functional claim limitations. The number of independent claims in the litigated patents was roughly 30 percent higher than in the control group. The number of words per independent claim was roughly 9 percent lower for the litigated patents. Finally, almost 27 percent of the litigated patents contained functional claim language, compared to only 21 percent in the control group.

Table 4-2 provides a comparison of the means of the patent- and examination-related variables across the GAO sample with the propensity-matched control group. Note that since we included small entity status, continuation history, and foreign origin among the predictors in the model to generate the propensity scores, we do not include them here.[47] As expected, none of their means is significantly different across the GAO sample and control group.

Table 4-2. Means Comparisons for Litigated Patents and Propensity Score-Matched Control Group

	Litigated Patents (n=975)		Control Group (n=975)	
Examiner Seniority	**Mean**	**Std Error**	**Mean**	**Std Error**
GS-12 and Lower*	0.217	0.013	0.251	0.014
GS-13	0.161	0.012	0.179	0.012
GS-14 and Higher**	0.568	0.016	0.518	0.016
GS Level Missing	0.053	0.007	0.051	0.007
Examination Variables				
Number of IDS Filings***	3.889	0.156	2.797	0.124
Number of Interviews***	0.419	0.027	0.275	0.021
Pendency Before Examiner***	1.735	0.047	1.556	0.041
Number of RCEs***	0.294	0.023	0.199	0.018
At Least One Appeal**	0.021	0.005	0.009	0.003
FAOM Allowance	0.136	0.011	0.203	0.013
Patent Claims Variables				
Number of Independent Claims***	4.117	0.162	3.276	0.105
Number of Words per Independent Claim***	137	3	154	5
Functional Claim	0.265	0.014	0.234	0.014

*** Indicates that the mean values are significantly different at the 1-percent level.
**Indicates that the mean values are significantly different at the 5-percent level.
* Indicates that the mean values are significantly different at the 10-percent level.

Some results differ from those using the three-year forward citations-matched control group, reported in Table 4-1:

- Applications resulting in litigated patents have longer examination pendency. Using the three-year forward citation control group, we found no significant effect of examination pendency. Using the propensity score control group, we find a stronger effect indicating that applications that result in litigated patents take a longer time from docketing to allowance than those in the control group. In the logistic analysis, we will see if this result is maintained when controlling for the other patent- and examination-related variables.
- Litigated patents have a higher number of RCE filings. The result here is significant at the 1-percent level. The result using the three-year

forward citation-matched control group was similar, but not as strong, and it was not statistically significant.

- Litigated patents are more likely to have had a decision on an appeal. An applicant may choose to file an appeal if any claim in the application has been rejected by the examiner in at least two Office actions. The results in Table 4-2 indicate that the rate of appeal for litigated patents, although small at 2 percent, is twice as high as that for patents in the control group. The result using the three-year forward citation-matched control group was similar, but not as strong, indicating that the rate of appeal for litigated patents is about 30 percent higher than for the control group. The result for the 3-year forward citations control group is not statistically significant (see Table 4-1).
- Litigated patents are not significantly more likely to contain functional claim language. This partially contradicts the finding reported in Table 4-1. Roughly 21 percent of the patents in the three-year forward citation-matched control group contained functional claim language while 23.4 percent of the patents in the propensity score-matched control group contained such language. This is still a smaller percentage than the 26.5 percent of the litigated patents with functional claiming, but the difference in this case is not great enough to be statistically significant.

4.1.2. Conditional Logistic Results

Although the means comparisons are illuminating, they do not control for any potential correlations among the patent- and examination-related variables of interest. The conditional logistic models do take these correlations into account. The results of these models are presented in Table 4-3. The first two columns report the results for the model using the three-year forward citation-matched control group (Model 1). The last two columns report the results for the model using the propensity score-matched control group (Model 2). Note that entity size, continuation history and foreign status are not controlled for in Model 2, because these variables were used to generate the propensity scores, which we used for matching purposes. For each model, we report the odds ratio and the z-statistic, which is used to determine whether the odds ratio is significantly different from one.[48]

Table 4-3. Conditional Logistic Model Results for Patent Infringement Litigation

	3-Year Forward Citations (Model 1)			Propensity Score (Model 2)		
	Odds Ratio		z-statistic	Odds Ratio		z-statistic
Examiner Seniority (Base: GS-14 and Higher)						
GS-12 and Lower	0.847		-1.22	0.739	**	-2.41
GS-13	0.776		-1.6	0.739	**	-2.18
GS Level Missing	0.687		-1.14	0.829		-0.54
Small Entity	2.717	***	7.02			
Continuation History Number of Domestic Parents	1.322	***	3.85			
Pendency from Earliest Parent to		***				
Docketing	1.380		4.75			
Foreign Applications						
Foreign Priority Claim	0.439	***	-5.08			
371 Case	1.027		0.08			
Examination Variables						
Number of IDS Filings	1.054	***	2.98	1.075	***	3.86
Number of Interviews	1.313	***	2.86	1.259	***	2.78
Pendency Before Examiner	0.928		-0.96	1.020		0.32
Number of RCEs	1.097		0.85	1.123		1.22
At Least One Appeal	1.922		1.42	2.244		1.61
FAOM Allowance	0.624	**	-2.31	0.605	***	-3.07
Application Year (1981-1990 is Comparison Group)						
1991-1995	0.837		-0.27	0.951		-0.06
1996-2000	1.364		0.4	0.738		-0.33
2001-2005	1.452		0.44	0.935		-0.07
2006-2011	1.857		0.66	1.339		0.28
Patent Claims Variables Number of Independent Claims	1.066	***	3.1	1.050	***	2.89
Number of Words per		***			***	
Independent Claim	0.998		-2.98	0.998		-3.56
Functional Claim	1.399	**	2.3	1.148		1.06

*** Result is statistically significant at the 1-percent level.

** Result is statistically significant at the 5-percent level.

* Result is statistically significant at the 10-percent level.

In Figure 4-1, we rank the patent- and examination-related variables based on the size of their impact on the likelihood of litigation.[49] Because we want to include measures of entity size, continuation history, and foreign priority in reporting the results, we rely on Model 1 for the generation of this figure. For those characteristics that are measured using indicator variables, the size of the impact is simply the odds ratio as reported in Table 4-3. We also include error bounds around the point estimates. We determined the size of the impact differently for the characteristics that we measured as continuous variables. For these we consider the impact of changing the value of the variable from the variable's 25th percentile to its 75th percentile. For instance, the bottom 25 percent of the patents in our analysis have 79 or fewer words per independent claim (the 25th percentile), while the top 25 percent have 194 or more words per independent claims (the 75th percentile). Thus, the impact that we calculate for this variable is the impact that the model predicts if we were to change the average number of words per independent claim from 79 to 194. The model predicts that this would decrease the odds of litigation by 20 percent.

Many of the results are consistent across the two models. For instance, the number of IDS filings and number of interviews are found to be positively and significantly related to the likelihood of litigation in both. These results are also consistent with the results of the means comparisons reported in Tables 4-1 and 4-2. Also, the results regarding the relationships between first-action allowance, the number of independent claims, and the number of words per independent claim are consistent across the two models.

Model 1 and Model 2 differ with regard to the extent of the impact of functional claiming and of examiner GS-level on the likelihood of litigation. In Model 1, the estimated impact of functional claim language is greater and statistically significant. In addition, examiner GS-level is not significant in Model 1, whereas it is in Model 2. Note, however that the results are consistent regarding the direction of the estimated effects of GS-level on litigation. Still, given their lack of robustness, the results on these factors should be approached with caution at this stage.

Another interesting result from Model 1 is that the highest-impact variables are all measures over which the USPTO has little or no control (see Figure 4-1). The four most important variables are all related to entity size, foreign origin, and continuation history. That said, while those variables represent applicant behavior, USPTO policy can certainly influence that behavior.[50] A logistic model containing just the entity size, foreign origin, and continuation history variables predicts litigation almost as well as Model 1, and far better than Model 1 would without the inclusion of these variables (i.e., with just the patent-examination variables included). In fact, the conditional logistic model including just the entity size, foreign origin, and continuation variables outperforms Model 1 based on the Bayesian information criterion (BIC), although it slightly underperforms it based on the Akaike information criterion (AIC).[51,52]

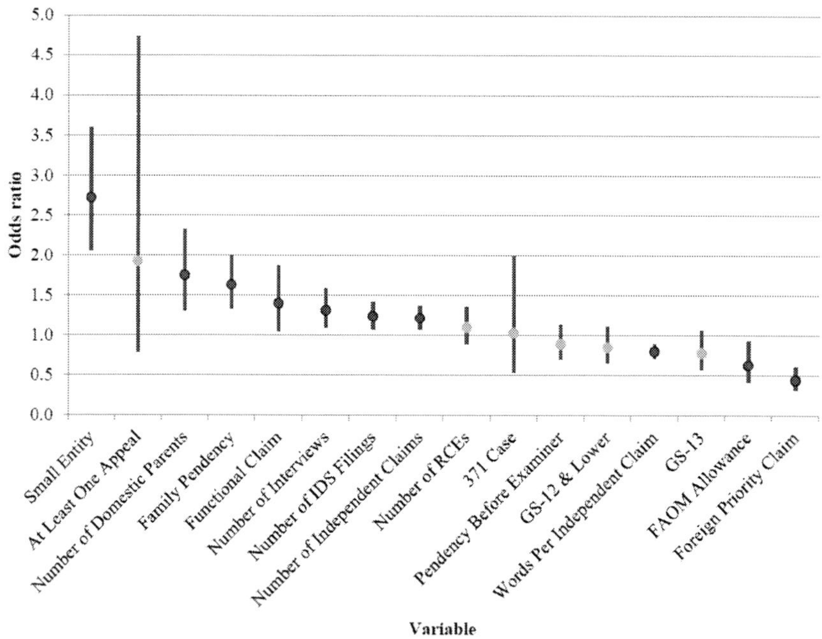

Figure 4-1. Ranking the Impact of the Explanatory Variables on Litigation (Model 1).

The upshot of this is that the examination-related variables only affect the likelihood of litigation at the margins. Even then, at least three of these variables (IDS filings, interviews, and first-action allowances) are driven as much by the applicants as by the USPTO. IDS filings and interviews may indicate greater applicant involvement in the examination process, which is likely to be a signal of a higher perceived value of the invention on the part of the applicant. Higher perceived value would translate into a greater probability of litigation, all other factors being equal. First-action allowances may signal patents whose applications came into the USPTO with well-drafted and/or relatively narrow claims. This could affect litigation in two ways. First, narrower claims are associated with less uncertainty. Second, applications with narrower claims may be less valuable.

As expected, patents with more independent claims are more likely to be litigated. This provides support to our expectation that the likelihood of infringement would be positively correlated with the number of independent claims. We also find that patents with more words per independent claim (after controlling for technology area in the matching) are less likely to be litigated. This makes sense because we expect that claims with more words have better-defined boundaries, leading to less uncertainty regarding these boundaries and thus lowering the likelihood of misunderstandings between patent holders and potential infringers.

4.2. The Likelihood of an IPR Petition

4.2.1. Means Comparisons

Table 4-4 provides a comparison of the means of the explanatory variables across the PTAB sample and the citation-matched control group.

Table 4-4. Means Comparisons for IPR Petitioned Patents and the Three-Year Citation Matched Control Group

	Petitioned Patents (n=1,014)		Control Group (n=1,014)	
Examiner Seniority	**Mean**	**Std Error**	**Mean**	**Std Error**
GS-12 and Lower***	0.202	0.013	0.283	0.014
GS-13	0.155	0.011	0.180	0.012
GS-14 and Higher***	0.605	0.015	0.501	0.016
GS Level Missing	0.038	0.006	0.036	0.006
Small Entity* **	0.293	0.014	0.159	0.011
Continuation History				
Number of Domestic Parents***	2.429	0.124	0.911	0.050
Pendency from Earliest Parent to Docketing***	2.129	0.065	1.280	0.045
Foreign Applications				
Foreign Priority Claim***	0.104	0.010	0.309	0.015
371 Case***	0.032	0.005	0.081	0.009
Examination Variables				
Number of IDS Filings***	5.654	0.279	3.954	0.185
Number of Interviews	0.356	0.028	0.319	0.022
Pendency Before Examiner***	1.873	0.053	2.168	0.049
Number of RCEs	0.394	0.026	0.371	0.024
At Least One Appeal	0.018	0.004	0.020	0.004

	Petitioned Patents (n=1,014)		Control Group (n=1,014)	
Examiner Seniority	**Mean**	**Std Error**	**Mean**	**Std Error**
FAOM Allowance	0.139	0.011	0.153	0.011
Patent Claims Variables				
Number of Independent Claims***	4.166	0.132	3.458	0.103
Number of Words per Independent Claim***	156	3	172	3
Functional Claim	0.201	0.013	0.193	0.012

*** Indicates that the mean values are significantly different at the 1-percent level.

The results of the means comparisons indicate the following.

- IPR-petitioned patents are less likely to have been allowed by examiners of GS-12 or below and more likely to have been allowed by examiners of GS-14 or above. Twenty percent of the petitioned patents were allowed by examiners of GS-12 or below compared to 28 percent of the patents in the control group. At the same time, 60 percent of the petitioned patents were allowed by examiners of GS-14 or above compared to 50 percent of the control group patents. These differences are both statistically significant at the 1- percent level.
- IPR-petitioned patents are more likely to have been granted to small entities. Roughly 30 percent of the petitioned patents were granted to small entities, while only 16 percent of the patents in the control group were granted to such entities. These results are quite similar to those we found for litigation filings.
- IPR-petitioned patents are more likely to have an extensive continuation history. The average number of domestic parents for IPR-petitioned patents is roughly 2.5 times the number for patents in the control group. Also, the average pendency from the filing date of the earliest parent to the docketing date of the later-issued application is 66 percent longer for IPR-petitioned patents.
- IPR-petitioned patents are less likely to have been granted on an application with a foreign or international priority. Either way that we measure foreign applications – a foreign priority claim or a 35 U.S.C. § 371 national stage entry – we get the same result. Only 10 percent of the IPR-petitioned patents had evidence of a foreign priority claim compared to 31 percent of the patents in the control group. Likewise, 3 percent of the IPR-petitioned patents were 35 U.S.C. § 371 national stage entries compared to 8 percent for the control group.

- IPR-petitioned patents have a higher number of IDS filings during examination.[53] The average number of IDS filings for IPR-petitioned patents was 42 percent higher than the average for the control group.
- IPR-petitioned patents are issued from applications that have shorter examination pendencies. The average pendency from docketing to allowance for IPR-petitioned patents was nearly 15 percent lower than for the control group.
- IPR-petitioned patents tend to have more independent claims and fewer words per independent claim. The number of independent claims in the IPR-petitioned patents was roughly 20 percent higher than in the control group. The number of words per independent claim was roughly 9 percent lower for the IPR-petitioned patents.

Table 4-5 provides a comparison of the means of the explanatory variables across the PTAB sample and the propensity-score-matched control group. Because we included small entity status, continuation history, and foreign origin among the predictors in the model to generate the propensity scores, we do not include them here. As expected, none of the means is significantly different across the PTAB sample and control group.

Table 4-5. Means Comparisons for IPR Petitioned Patents and the Propensity Score-Matched Control Group

	Petitioned Patents (n=1,014)		**Control Group** (n=1,014)	
	Mean	**Std Error**	**Mean**	**Std Error**
Examiner Seniority				
GS-12 and Lower**	0.202	0.013	0.246	0.014
GS-13	0.155	0.011	0.177	0.012
GS-14 and Higher***	0.605	0.015	0.546	0.016
GS Level Missing	0.038	0.006	0.032	0.005
Examination Variables				
Number of IDS Filings***	5.654	0.279	4.130	0.224
Number of Interviews**	0.356	0.028	0.282	0.021
Pendency Before Examiner**	1.873	0.053	1.707	0.043
Number of RCEs***	0.394	0.026	0.272	0.020
At Least One Appeal*	0.018	0.004	0.009	0.003
FAOM Allowance	0.139	0.011	0.159	0.011
Patent Claims Variables				
Number of Independent Claims***	4.166	0.132	3.488	0.099

	Petitioned Patents		Control Group	
	(n=1,014)		(n=1,014)	
	Mean	**Std Error**	**Mean**	**Std Error**
Number of Words per Independent Claim***	156	3	170	3
Functional Claim	0.201	0.013	0.187	0.012

*** Indicates that the mean values are significantly different at the 1-percent level.
** Indicates that the mean values are significantly different at the 5-percent level.
* Indicates that the mean values are significantly different at the 10-percent level.

Some results differ from those using the citations-matched control group, reported in Table 4-4.

- IPR-petitioned patents have more examiner interviews during examination. When using the propensity score-matched sample, we find a statistically significant relationship between the number of examiner interviews and the filing of an IPR petition. In fact the number of interviews for the IPR-petitioned patents is roughly 25 percent higher than for the control group. The result is not much different from the one we found using the forward citation-matched control group, but is slightly larger and thus meets the requirement for statistical significance.
- IPR-petitioned patents have longer examination pendency. When using the propensity score-matched sample we find that IPR-petitioned patents have slightly longer examination pendency than those in the control group. This result is not consistent with the one we found using the forward citation-matched sample.
- IPR-petitioned patents have a higher number of RCE filings. The result here is significant at the 1-percent level. The result using the three-year forward citation-matched control group was in the same direction but much smaller, and it was not statistically significant.
- IPR-petitioned patents are more likely to have had at least one completed appeal. The results in Table 4-5 indicate that the rate of appeal for IPR-petitioned patents, although small at 2 percent, is twice as high as that for patents in the control group. The result using the three-year forward citation-matched control group was non-existent (see Table 4-4).

4.2.2. Conditional Logistic Results

The results of the conditional logistic models are presented in Table 4-6. The first two columns report the results for the model using the citation-matched control group (Model 3). The last two columns report the results for the model using the propensity-score-matched control group (Model 4). As was the case with the litigation models, entity size, continuation history, and foreign status are not controlled for in Model 4, because these variables were used to generate the match. For each model, we report the odds ratio and the z-statistic, which is used to determine whether the odds ratio is significantly different from one.

In Figure 4-2, we rank the significant explanatory variables based on the size of their impact on the likelihood of a petition for IPR. Because we want to include measures of entity size, continuation history, and foreign priority in reporting the results, we rely on Model 3 for the generation of this figure. For those characteristics that are measured using indicator variables, the size of the impact is simply the odds ratio as reported in Table 4-6. We also include error bounds around the point estimates. We determined the size of the impact differently for the characteristics that we measured as continuous variables. For these we consider the impact of changing the value of the variable from the variables 25th percentile to its 75th percentile. For instance, 25 percent of the patents in our analysis have 105 or fewer words per independent claim (the 25th percentile), while 25 percent have 207 or more words per independent claim (the 75th percentile). Thus, if we were to change the average number of words per independent claim from 105 to 207 the model predicts that the odds of an IPR petition would decrease by 23 percent.

Table 4-6. Conditional Logistic Model Results for IPR-Petitioned Patents

	3-Year Forward Citations (Model 3)			**Propensity Score (Model 4)**		
	Odds Ratio		**z-statistic**	**Odds Ratio**		**z-statistic**
Examiner Seniority (Base: GS-14 and Higher)						
GS-12 and Lower	0.681	**	-2.57	0.716	***	-2.69
GS-13	0.722	**	-1.96	0.777	*	-1.86
GS Level Missing	1.079		0.26	1.080		0.27
Small Entity	2.485	***	6.45			
Continuation History						
Number of Domestic Parents	1.279	***	4.95			

	3-Year Forward Citations (Model 3)			Propensity Score (Model 4)		
	Odds Ratio		**z-statistic**	**Odds Ratio**		**z-statistic**
Pendency from Earliest Parent to						
Docketing	1.133	***	2.57			
Foreign Applications						
Foreign Priority Claim	0.338	***	-6.61			
371 Case	0.790		-0.78			
Examination Variables						
Number of IDS Filings	1.052	**	2.13	1.031	*	1.71
Number of Interviews	1.094		1.02	1.093		1.34
Pendency Before Examiner	0.848	***	-2.68	0.997		-0.06
Number of RCEs	1.144		1.49	1.181	**	1.96
At Least One Appeal	1.265		0.53	1.890		1.2
FAOM Allowance	0.896		-0.69	1.025		0.18
Application Year (1981-1995 is Comparison Group)						
1996-2000	1.197		0.38	1.048		0.12
2001-2005	1.197		0.33	1.017		0.04
2006-2013	1.185		0.26	0.988		-0.02
Patent Claims Variables						
Number of Independent Claims	1.066	***	3.41	1.062	***	3.51
Number of Words per						
Independent Claim	0.997	***	-3.39	0.999	***	-2.89
Functional Claim	1.116		0.72	1.062		0.47

*** Result is statistically significant at the 1-percent level.
** Result is statistically significant at the 5-percent level.
* Result is statistically significant at the 10-percent level.

For the most part, the results are quite robust across the two models. For instance, in each case we find a monotonic positive relationship between the GS-level of the allowing examiner and the likelihood of filing a petition for IPR. For instance, the odds that a patent allowed by an examiner with GS-level of 12 or lower will be petitioned are roughly 30 percent lower than the odds that a patent allowed by an examiner with GS-14 or higher will be. The odds that a patent that was allowed by a GS-13 examiner will be petitioned are somewhere in the middle, but are closer to the odds for the examiners at the GS-12 or lower levels. The results on the relationship between IDS filings and the filing of an IPR petition are consistent across the two models, although the relationship is somewhat stronger in Model 3, with a higher odds ratio and higher level of statistical significance. The results across the models are also consistent when it comes to the claims variables. The odds ratios for both the number of independent claims and words per independent claim variables are nearly identical across the two models and they are all significant at the 1-percent level.

The only inconsistencies between models lie with the examination pendency and RCE variables. The examination pendency variable is significant in Model 3, but not in Model 4. This is similar to the results from the comparison of means (Table 4-4), where longer pendency was found to be associated with a lower likelihood of IPR petition. The number of RCEs filed during examination is marginally significant in Model 4. The results indicate that the filing of each additional RCE increases the odds of an IPR petition being filed by 18 percent. The RCE variable is not significant in Model 3.

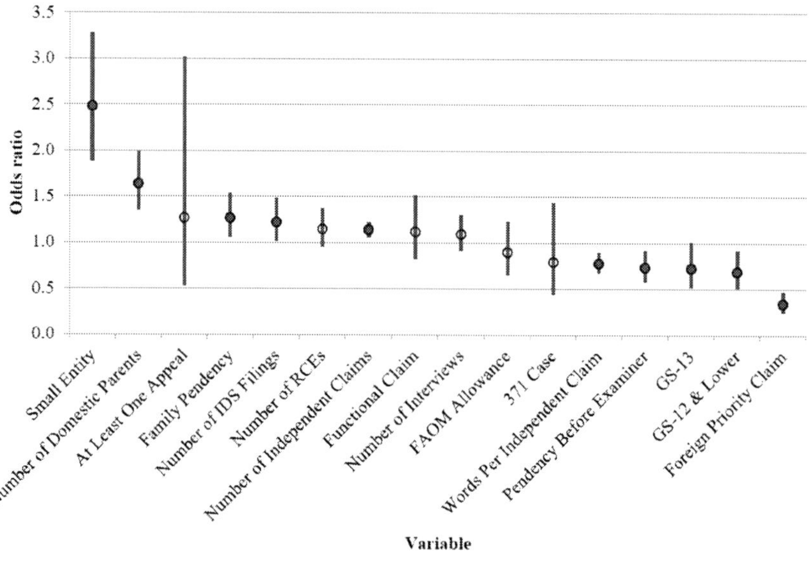

Figure 4-2. Ranking the Impact of the Explanatory Variables on IPR (Model 3).

As in our analyses of litigation incidence, we find that the highest-impact variables tend to be measures over which the USPTO has little control (see Figure 4-2). Once again, the most important variables are all related to entity size, foreign origin, and continuation history. A logistic model containing just the entity size, foreign origin, and continuation history variables predicts litigation almost as well as Model 3, and far better than Model 3 would without the inclusion of these variables (i.e., with just the patent-examination variables included). In fact, the conditional logistic model including just the entity size, foreign origin, and continuation variables outperforms Model 3 based on the Bayesian information criterion (BIC), although it slightly underperforms it based on the Akaike information criterion (AIC).[54]

5. Comparison of Instituted and Non-Instituted IPR-Petitioned Patents

In addition to modeling the incidence of IPR petitions as discussed above, we also investigate the relationship between the same set of patent examination characteristics and the likelihood that the PTAB had instituted an IPR proceeding for at least one claim that had been petitioned for review. In this case the sample consists of all IPR-petitioned patents for which the PTAB had issued a ruling on whether to institute an IPR proceeding by the middle of July 2014. In section 6, we also consider the written decisions for those petitions that were instituted.

5.1. Data and Methods

In our analysis of instituted vs. non-instituted proceedings for IPR-petitioned patents, we do not need to use a matched case-control design. In this case, our population of interest includes only those patents that had been challenged by filing a petition for an IPR proceeding. Thus, we start with our list of patents from the PTAB and limit our analysis to those patents for which an institution decision has been made. Our dependent variable was equal to one if the proceeding had been instituted for at least one claim and equal to zero otherwise.

As we described above, a logistic model is a good choice when modeling the likelihood that an event occurs. In this case the event is the institution by the PTAB of an IPR proceeding involving at least one claim. We included the same patent- and patent examination-related variables as we did in the matched case-control studies. However, in this case we use an unconditional logistic model, because we do not need to construct a matched control group for the IPR-petitioned patents for which the proceeding was instituted. The control group consists of those IPRpetitioned patents for which the PTAB decided not to institute the proceeding as to any claim. The results of an unconditional logistic model can be interpreted in a similar fashion to those of a conditional logistic model. We report odds ratios and measures of statistical significance.

5.2. Results

5.2.1. Descriptive Statistics

Table 5-1 presents the means comparisons between the IPR-petitioned patents for which the proceeding was instituted for at least one claim and those IPR-petitioned patents for which the proceeding was not instituted. Generally, we find no significant differences between the two groups. There are a couple of exceptions and they both relate to the pay grade of the allowing examiner. Roughly 60 percent of the IPR-petitioned patents had been allowed by examiners with pay grades at the GS-14 level or higher. There is no significant difference in the proportion allowed by these GS-14 or higher examiners. Where we find a significant difference is in the mix of examiners with pay grades below the GS-14 level. For those IPR-petitioned patents for which the proceeding was instituted as to at least one claim, roughly 25 percent of the allowing examiners had pay grades below the GS-13 level and roughly 15 percent had their pay grade at the GS-13 level. This is reversed for the IPR-petitioned patents for which the proceeding was not instituted. Roughly 15 percent of the allowing examiners had pay grades below the GS-13 level and roughly 25 percent had their pay grade at the GS-13 level.

Table 5-1. Means Comparisons for Instituted vs. Non-Instituted IPR-Petitioned Patents

	Instituted (n=486)		**Not Instituted** (n=159)	
	Mean	**Std Error**	**Mean**	**Std Error**
Examiner Seniority				
GS-12 and Lower**	0.235	0.019	0.157	0.029
GS-13**	0.130	0.015	0.239	0.034
GS-14 and Higher	0.609	0.022	0.572	0.039
GS Level Missing	0.027	0.007	0.031	0.014
Small Entity	0.276	0.020	0.314	0.037
Continuation History				
Number of Domestic Parents	2.706	0.210	2.767	0.373

	Instituted		Not Instituted	
	(n=486)		(n=159)	
	Mean	Std Error	Mean	Std Error
Pendency from Earliest Parent to Docketing	2.131	0.094	2.255	0.161
Foreign Applications				
Foreign Priority Claim	0.097	0.013	0.069	0.020
371 Case	0.031	0.008	0.025	0.012
Examination Variables				
Number of IDS Filings	6.126	0.508	4.937	0.420
Number of Interviews	0.356	0.036	0.472	0.097
Pendency Before Examiner	1.897	0.082	1.915	0.135
Number of RCEs	0.370	0.036	0.465	0.090
At Least One Appeal	0.019	0.006	0.019	0.011
FAOM Allowance	0.142	0.016	0.145	0.028
Patent Claims Variables				
Number of Independent Claims	4.403	0.227	4.031	0.281
Number of Words per Independent Claim	156	4	156	7
Functional Claim	0.200	0.018	0.157	0.029

** Indicates that the mean values are significantly different at the 5-percent level.

5.2.2. Logistic Results

Table 5-2 presents the results of the logistic regression.[55] As was the case with the comparisons of means, we find that most of the explanatory variables have no significant relationship with the decisions regarding the institution of IPR proceedings. We do find that patents that were allowed by examiners whose pay grade was GS-13 at the time of allowance are less likely to have been instituted. We also find a very small and only marginally significant positive effect of IDS filings. The one result that truly stands out is the relationship between the year of issue and the institution of claims. We find that patents that were issued roughly a decade ago (in the early 2000s) are much more likely to have had the IPR proceeding instituted by the PTAB. This result might be confounded somewhat by the inclusion of the application year variables. Note that many of the patents that were issued in the early 2000s would have application dates from the late 1990s. The impact of an application date in the late 1990s, according to our results, is a decrease in the likelihood of institution. Thus, these two effects, opposite in sign, may be a result of a problem with the exact model we are estimating here. However, when we drop the application year variables, the results for the patent issue year variables, although mitigated somewhat, continue to tell the same story: among petitioned patents, those issued in the early 2000s are by far the most likely to have had an IPR proceeding instituted.

This result may, at least in part, be reflective of the Supreme Court's 2007 decision in KSR v. Teleflex, 550 U.S. 398 (2007). The KSR decision is widely seen as having relaxed the criteria for finding a claim to be unpatentable due to obviousness. Although the Court reiterated that a reasoned explanation for modifying the prior art to arrive at a claimed invention was required for a proper determination of obviousness, it moved away from the idea that an explicit teaching suggestion, or motivation to modify had to be provided by the prior art itself.[56] Thus, claims issued in the early 2000s – which may not have been considered obvious under a pre-KSR understanding of the law – could have been determined to be unpatentable in an IPR proceeding in view of the current understanding of the law of obviousness.

Table 5-2. Logistic Model Results for Instituted vs. Non-Instituted IPR-Petitioned Patents

	Odds Ratio		z-statistic
Examiner Seniority (Base: GS-14 and Higher)			
GS-12 and Lower	1.504		1.530
GS-13	0.553	**	-2.400
GS Level Missing	0.920		-0.150
Small Entity	0.822		-0.890
Continuation History			
Number of Domestic Parents	0.993		-0.330
Pendency from Earliest Parent to Docketing	0.970		-0.570
Foreign Applications			
Foreign Priority Claim	1.487		1.010
371 Case	1.095		0.140
Examination Variables			
Number of IDS Filings	1.039	*	1.840
Number of Interviews	0.875		-1.170
Pendency Before Examiner	0.956		-0.530
Number of RCEs	0.880		-1.060
At Least One Appeal	0.903		-0.130
FAOM Allowance	1.041		0.140
Application Year (1981-1995 is Comparison Group)			
1996-2000	0.427	*	-1.780
2001-2005	0.655		-0.700
2006-2011	0.475		-0.980
Patent Claims Variables			
Number of Independent Claims	1.008		0.330
Number of Words per Independent Claim	0.999		-0.610

	Odds Ratio		z-statistic
Functional Claim	1.517		1.560
Number of 3-Year Forward Citations	1.000		0.010
Patent Issue Year (1995-1999 is the Comparison Group)			
2000-2004	3.359	***	2.890
2005-2009	1.957		1.210
2010-2013	2.577		1.330

*** Result is statistically significant at the 1-percent level.
** Result is statistically significant at the 5-percent level.
* Result is statistically significant at the 10-percent level.

6. Hands-on Review of Patents Which Have Undergone IPR

6.1. Data and Methods

As part of our investigation in response to the GAO Report, we conducted a hands-on review of the prosecution history of all patents for which the PTAB had instituted an IPR proceeding and issued a final written decision (FWD) as of mid-July 2014. Although the PTAB granted 89 petitions for IPR during this time frame, there were only 78 patents in this group because 11 patents had been the subject of more than one petition. The issue dates for these patents were between 1996 and 2012, as shown in Figure 6-1. A wide variety of technologies, and 45 different patent owners, were represented. Of these 78 patents, 64 of them (82 percent) had at least one claim canceled in view of the PTAB's determination of unpatentability over the prior art. Some summary statistics about the patents are included in Table 6-1, and Appendix I includes detailed information about the cases.

There were 21 reviewers who undertook a detailed investigation of the 78 patents in the IPR study group. The reviewers considered the prosecution history of the patent application before the patent examiner, the petition for IPR submitted by the third party, the PTAB's order instituting IPR, and the PTAB's FWD. All of these reviewers were experienced Patent Quality Assurance Specialists with the Office of Patent Quality Assurance (OPQA). Although the reviewers do not assess cases that have undergone IPR as part of their ordinary duties, we found them to be particularly suited to this task. They regularly evaluate the work of patent examiners both while an application is being prosecuted and after it has been allowed, with an eye toward adequacy of the prior art search, appropriate evaluation of potential prior art references, and proper drafting of anticipation and obviousness rejections. In the case of allowed applications, the reviewers also consider the correctness of the examiner's decision to allow. Thus, this group of reviewers was well-versed in the issues raised in IPR proceedings.

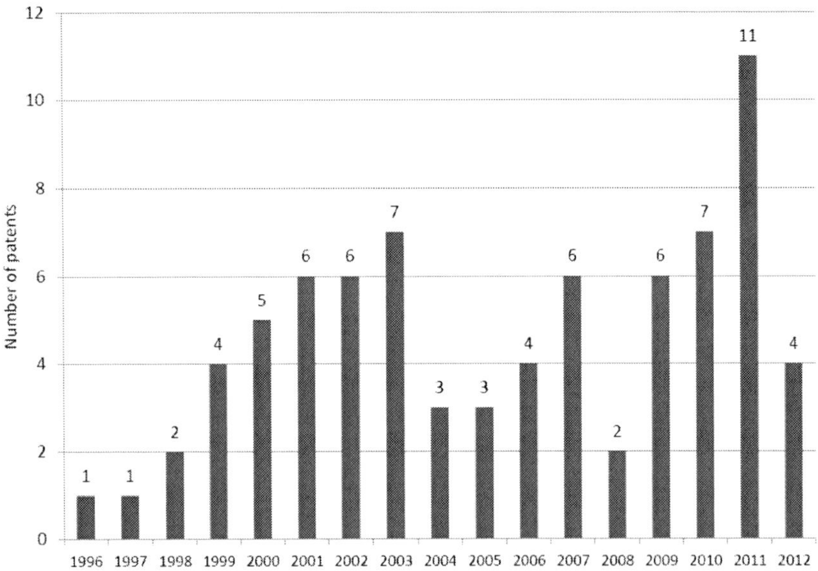

Figure 6-1. Distribution of Patents in Hands-On Study by Issue Date.

Table 6-1. Basic Information about the 78 Patents in Hands-On Study

# (%) of Cases Reviewed by Discipline	Chemical	19 (24%)
	Electrical	45 (48%)
	Designs	1 (1%)
	Mechanical	13 (17%)
Characteristics	# of Unique Patent Owners	45
	Average # of Claims in Petitioned Patents	32
	Average # of Claims Found Unpatentable	9
	# (%) Case with at Least OneClaim Found Unpatentable	64 (82%)
	# (%) Case with All Claims Found Unpatentable	19 (24%)

A questionnaire for the reviewers' use was developed in consultation with OPQA managers (see Appendix II). The questionnaire was designed to gather information about six aspects of the prosecution history:

A. sources of prior art relied on in the PTAB's decision,
B. the examiner's prior art search,
C. the examiner's handling of prior art rejections,
D. the examiner's handling of application continuity,
E. substantive interviews between the examiner and the applicant, and
F. claim interpretation issues.

Reviewers were assigned to work on prosecution histories in areas of their technical competence. A single reviewer completed all facets of the review of the prosecution history for each patent. In addition to asking specific questions about the prosecution history, the questionnaire prompted the reviewer to comment, when relevant, about reasons why the examiner may have reached a different conclusion from the PTAB on patentability of the claims.

The data obtained from this hands-on review must be considered preliminary. We recognize that the small sample size, dictated by the recent inception of IPR, significantly limits the ability to draw meaningful conclusions from the study. For example, certain subcategories identified on the questionnaire included only one or two patents, so identifying trends was not possible. Time constraints mandated that each case was reviewed by a single reviewer, and that several reviewers were required from each technological specialty. That is, we were unable to assign all of the proceedings in a particular technology to the same reviewer. If we had been able to rely on a smaller number of expert reviewers, it may have ensured a more uniform application of the reviewing criteria. Although we did conduct a trial run of the questionnaire on four IPR proceedings, and we revised the questionnaire as appeared necessary before conducting the actual study, the desirability of additional revisions became apparent only after we were able to see the results of the entire study. Revision of the questionnaire to enhance its clarity and to further define the requested information would be desirable if this study were to be expanded. Hindsight bias on the part of the reviewers was also a possibility in our study. In order to provide information about the prosecution history as compared with the IPR proceeding, reviewers were necessarily aware that the patent had undergone IPR. If more time had been available, an independent review of the same prosecution histories by reviewers who were blind to the fact that these patents had undergone IPR may have reduced the hindsight bias and yielded more reliable information. Despite the limitations, the results serve to indicate potentially fruitful areas for further investigation to enhance patent quality.

6.2. Results

6.2.1. Sources of Prior Art Relied on in the Decision

Section I of the questionnaire related to the sources of prior art relied on in the PTAB decision. The PTAB determined that at least one claim was unpatentable in view of prior art in 64 out of the 78 patents (82 percent) in the study group. Of these, there were 38 patents (59 percent) in which the prior art relied on by the PTAB for the unpatentability decision had been available to the examiner during prosecution of the application. Prior art supplied by the applicant in an IDS accounted for 33 of the 38 patents, while the examiner supplied the prior art in the other five (see Figure 6-2). Also, it should be noted that changes in the law may have contributed to the differences observed between the examiner's patentability determinations and those of the PTAB. The 38 patents included some instances (29 instances or 45 percent) in which the examiner had never made a rejection over the art applied by the PTAB, and others (13 instances or 20 percent) in which the examiner had made a rejection over the art applied by the PTAB, but had later withdrawn the rejection. These two categories overlap; in other words, there were instances in which both types of prior art were relied upon by the PTAB. In 20 out of 64 cases (31 percent), there was no prior art applied during prosecution even though the PTAB found at least one claim to be unpatentable over prior art. In 61 of 64 cases (95 percent), prior art supplied by the third party petitioner was used as the basis of an unpatentability determination by the PTAB; a breakdown of this prior art by type (U.S. or foreign patent document or non-patent literature) is shown in Table 6-2.

The sources of prior art used by the PTAB as the basis of an unpatentability determination are shown in Figure 6-3. The results indicate that patent examiners sometimes fail to apply appropriate prior art, even though it is of record during prosecution. There were 38 cases in which the PTAB determined that at least one claim was unpatentable over prior art that was of record in the original prosecution. Figure 6-4 is a further refinement of the information conveyed in Figure 6-2. The same 38 applications are the subject of both figures. Figure 6-4 subdivides the 33 cases in which prior art references used by the PTAB as a basis for unpatentability had been cited on an IDS. In 4 cases, the examiner had based a rejection on some of the references but did not use others. In 6 cases, the examiner had based a rejection on a reference, but later withdrew the rejection. Finally, in 23 cases, the examiner had not based a rejection on a reference relied on by PTAB, even though it had been presented on an IDS during prosecution. Prior art provided by the applicant is usually cited on an IDS, and may also appear in the background section of the application.[57]

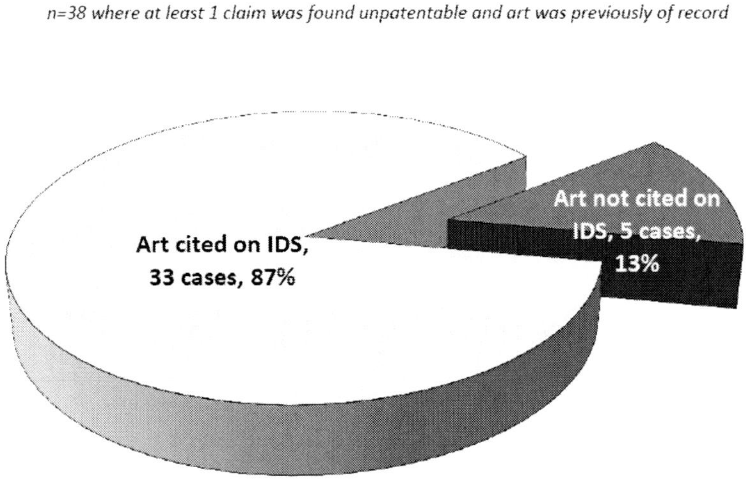

Figure 6-2. Source of Art Previously of Record Used by PTAB as Basis for Unpatentability Finding.

Table 6-2. Prior Art Provided by Third Party and Found to Render Claim Unpatentable

Prior art identified during IPR and determined by PTAB to render claims unpatentable:		Count	% of Total
US Patent or Application Publication	Yes	51	83.6
	No	10	16.4
Foreign Patent Document	Yes	26	42.6
	No	35	57.4
Non-Patent Literature	Yes	20	32.8
	No	41	67.2

n=61 where at least 1 claim was found unpatentable and new art was identified during IPR

We consider three possible reasons for the examiner's failure to rely on prior art cited on an IDS, even though the PTAB later found that art to render at least one claim unpatentable:

- The examiner simply neglects to consider prior art provided by the applicant.
- The examiner fails to understand the information in the reference or its relevance to the claimed invention.
- The examiner does not have time to do an adequate evaluation of the prior art provided by the applicant.

Simple neglect may be the easiest to address of these three possible reasons for an examiner's failure to apply prior art that is present in the application. Some examiners might acknowledge citations on an IDS by adding their initials in a pro forma way, without actually analyzing how the information cited could be used in a rejection of the pending claims. It is possible that such examiners misunderstand the applicant's duty to disclose information material to patentability under 37 CFR 1.56, and therefore wrongly believe that it is unlikely that they will find useful prior art cited on an IDS.

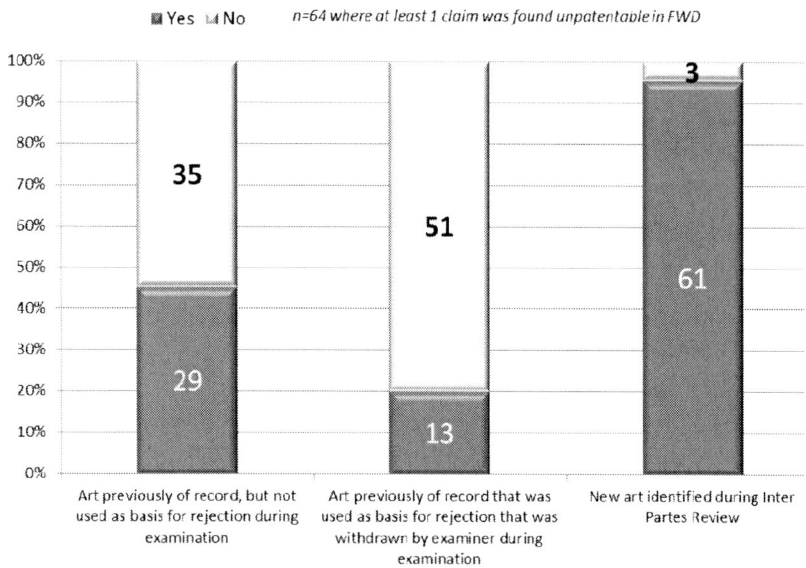

Figure 6-3. Sources of Prior Art Used by PTAB as Basis for Unpatentability.

If simple neglect is determined to be an issue, the Office could address it by reminding examiners of their duty to "[c]onsider the information properly submitted in an IDS in the same manner that the examiner considers other documents in Office search files while conducting a search of the prior art in a proper field of search."[58] It could also be worthwhile to educate examiners about the duty to disclose that falls on applicants and their attorneys, including the fact that an attorney who shirks this duty risks sanctions from the USPTO and possibly the relevant state bar. Finally, sharing the results of this study with examiners may be useful in helping them to recognize that applicant citations are a fruitful source of relevant prior art.

If it is determined that a lack of sufficient understanding of the relevant technology lies at the root of examiners' failure to apply appropriate prior art references, enhanced technical education may be helpful. The Office already provides some opportunities for technical education for examiners. With supervisory approval, examiners may use a limited number of non-examining hours to attend technical lectures of their own choosing. There are occasional educational visits for examiners to private or government laboratories or to industrial sites. Experts in various technical fields are also invited to speak to examiners at events such as "Tech Fairs" hosted by the various TCs. During the period in which the subject patents were granted, technical training was not always scheduled on a regular basis, and attendance ordinarily was voluntary. In furtherance of a White House executive action, the USPTO is currently "taking steps to make it easier for experts from industry and academia to provide relevant technical training to examiners by building upon existing programs and making the four regional satellite offices permanent."[59] Technical education efforts are being expanded and promoted to examiners as important ways to stay current in the technology relevant to their assigned dockets.

Insufficient time for examiners to perform an adequate review of all materials cited by an applicant may be a significant factor contributing to a failure to apply prior art references that are already of record. Although the current data set was not large enough to conclusively determine if a correlation exists between the number and/or length of the references cited and the incidence of failure to make a rejection, the numbers of references cited in the patents in our study sample (see Table 6-3) would suggest that examiner time constraints may be a factor. Further studies would be needed to gather more information. Although automated USPTO data searches can be conducted to count the number of IDS forms submitted, there is currently no automated way to count the number of items on each IDS, or to assess the length of each of the items.[60] Furthermore, it is generally thought that applicants in certain technologies tend to cite more items on IDS forms than applicants in other technologies. Therefore, it would be useful for further studies to include enough cases so that comparisons among the various technologies could be made. The Office is already considering the possibility of allowing more examiner time per case. If such a change is implemented, it would be worthwhile to stress the importance of using the additional time to ensure a thorough assessment of prior art provided by the applicant.

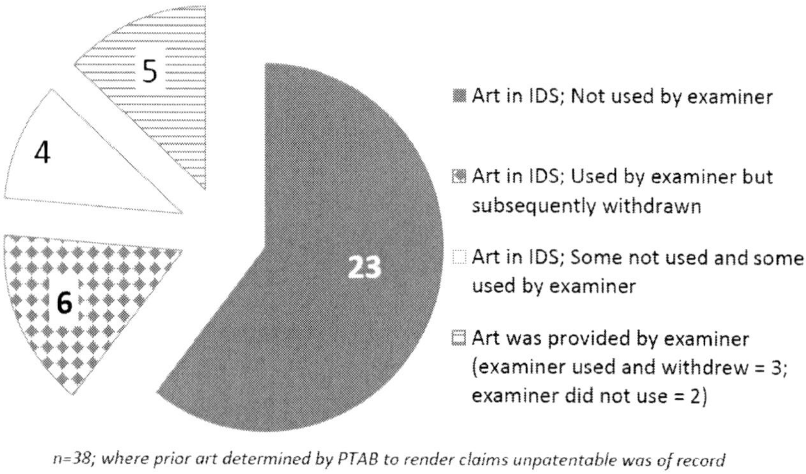

Figure 6-4. Art Previously of Record Used by PTAB as Basis for Finding of Unpatentability: Breakdown for References Cited on IDS.

If additional studies reveal that citation of unduly large numbers of references by applicants contribute to examiners' failure to apply prior art although it is present in the case, the applicants' role in the failure should also be investigated. If a reference is not material to patentability, it need not be disclosed to the examiner. Nevertheless, it is common for applicants to include references on an IDS out of an abundance of caution, in order to avoid any possible charge of inequitable conduct for failure to cite them. Some commentators have addressed the issue of "burying" material information among large numbers of prior art documents of questionable materiality.[61] In the present small-scale study, one application included 2,134 citations on IDS forms. When this outlier is excluded from the data, the average number of references cited on an IDS is 128 when the PTAB determined that at least one claim was unpatentable. In 2006, the Office proposed a rule change that would have discouraged citation of more than twenty documents, or of documents more than twenty-five pages long, but the change was never implemented and the proposal was withdrawn in 2010.[62] It may be appropriate to revisit the idea of addressing the issue of unduly large IDS citations through regulatory means.

Table 6-3. References Cited by Applicant on an IDS

	Claims Found Unpatentable and References Were Cited in IDS	**Claims Found Unpatentable and References Were Not Cited in IDS**
Valid N*	33	5
Average # References Cited**	128	46
Median # of References Cited **	39	15

* At least 1 claim was found unpatentable and art was previously of record;
** 1 outlier case with 2134 references cited has been excluded from summary statistics

Lastly, it is worth repeating that patents that end up being the subject of an IPR decision are not ordinary patents. They are subject to significant self-selection, and thus are not representative of the average patent. Furthermore, IPR-petitioned patents are typically also the subject of litigation. Thus, they are high-value patents for which there is a greater-than-average economic value associated with avoidance of examination-related errors. As a consequence, care must be used in drawing any broad inferences about examination practices more from this limited, self-selected, pool of patents.

6.2.2. The Examiner's Prior Art Search

Section II of the questionnaire relates to the examiner's prior art search. In the opinion of the reviewers, the examiner's search was deficient in more than half (37 prosecution histories or 61 percent) of the cases for which the PTAB determined that at least one claim was unpatentable over prior art that was not found by the examiner's search. The reviewers were not asked to conduct an independent prior art search. The result suggests an examiner search proficiency that is unacceptably low, however several caveats should be kept in mind. First, the patents in question are subject to significant self-selection: they were petitioned for an IPR based on prior art not found by the examiner, the proceeding was instituted by PTAB, and at least one claim was found to be unpatentable in view of the newly submitted prior art. The self-selection of these patents makes it inappropriate to conclude that this figure applies to the average patent examination. Additionally, the reviewers may be subject to a hindsight bias. That is, reviewers may have labeled a search "deficient" merely because a third party provided applicable prior art to the PTAB that was not located by the examiner, regardless of whether the examiner could reasonably have been expected to find that prior art by using the tools available at the time. It is also possible that a confirmation bias on the part of the reviewers, who knew that the cases they were reviewing contained claims that were ultimately found to be unpatentable by the PTAB, contributed to the reported low search proficiency. In fact, data from normal OPQA compliance reviews across technology centers and over several years indicates that in 90-95% of the cases, the reviewer does not uncover better prior art than that applied by the examiner. These data support the idea that it may not be appropriate to take the reported low search proficiency at face value. It would be useful to expand the study to include an additional review of the IPR cases by reviewers who were blind to their IPR status, and who conducted their own prior art search as is done for general compliance review. The results of such a study would shed more light on the question of adequacy of the examiner's search. However, conducting such a reviewer-blind investigation would present additional challenges. Although reviewers routinely evaluate allowed applications, they do not assess issued patents in the regular course of their work. It would be difficult to ensure that reviewers remain unaware that they were dealing with issued patents; that fact alone would suggest that patentability had been called into question in some way.

In hindsight, we recognize that question II.B., which asks reviewers to assess the possible reasons for finding a search to have been deficient (see Figure 6-5) is not well-designed. For example, a reviewer could deem a search to be "too limiting" for any of a variety of reasons, including failure to consult appropriate databases; failure to employ appropriate search terms, including technical terms, controlled vocabulary, and inventor information; failure to use truncation symbols, proximity operators, and Boolean operators effectively; and, failure to review the relevant sections of the patent classification system.[63] In other words, the choice "search is too limiting" could be inclusive of many different search defects, including those that question II.B. lists as potential alternatives. A more focused question would allow more meaningful conclusions to be drawn from the reviewers' assessments.

There is a sense in which every petitioned case can be considered to have been "deficient" in terms of the art considered – at least in the view of the petitioner. Every IPR petition requires submission of prior art. An IPR petition would not have been filed, and if it were, no proceeding would have been instituted, unless there was reason to believe that the art rendered at least one claim unpatentable. The question asked of the OPQA reviewers on the questionnaire was, "If prior art determined to render the claims unpatentable was not uncovered by examiner's search, was the search performed considered to be deficient?" Because we are considering a pool of cases for which relevant art was not found by the examiner, one could say that we are dealing with a self-selected group of "deficient" cases. The self-selection of instituted IPR cases anticipates a higher level of deficient search results relative to the average patent or non-instituted IPR petition. For this reason, it is not clear that the results about the adequacy of the search for this small sample set are an adequate reflection of the search proficiency of examiners in general.

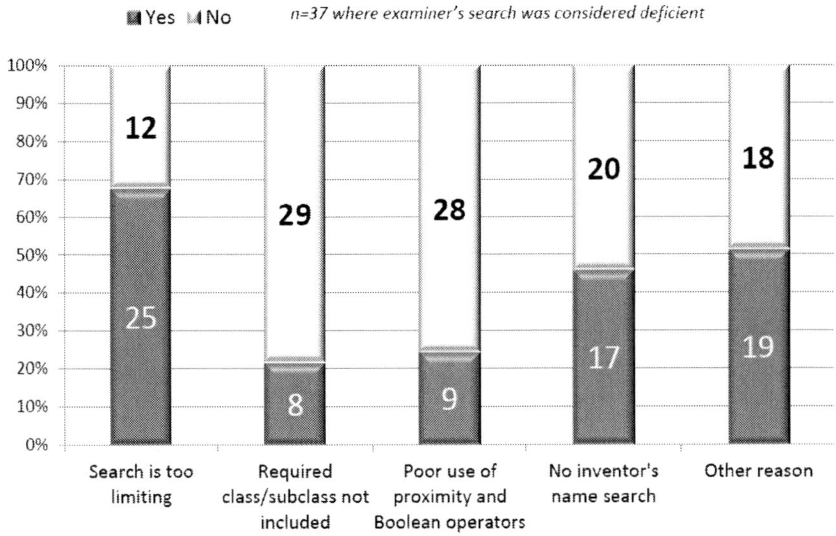

Figure 6-5. Reasons that Examiner's Search Is Considered Deficient.

Despite the limitations in the present methodology, it must be stated that enabling examiners to do the best job possible of finding prior art is an important goal of the Office. The Scientific and Technical Information Center (STIC), which has as its mission "to enhance patent quality and examiner effectiveness through search, information services, and training,"[64] is an important Office resource for examiners. The Office should work to ensure that STIC can continue to offer up-to-date services, and that examiners are aware of these services and can fully utilize them. Examiner participation in training opportunities offered by various database providers should also be encouraged and expanded. The AIA provided for pre-issuance submission of prior art by third parties, which can be another valuable means for examiners to identify relevant references.[65] The Office is exploring other innovative means of finding prior art, such as crowdsourcing,[66] and will continue to look for ways to ensure that the best prior art is available to the examiner at the earliest possible time.

6.2.3. The Examiner's Handling of Prior Art Rejections

Section III of the questionnaire dealt with the examiner's handling of prior art rejections. The data appear to indicate that when examiners choose to rely on a reference as the basis of a prior art rejection, they most often do so properly. Examiners usually make appropriate decisions regarding withdrawal of rejections. Out of the 78 IPR proceedings in the study group, there were only four (5 percent) in which the PTAB determined that a claim was unpatentable based on a prior art rejection that had been made by the examiner but later withdrawn (Figure 6-6). Furthermore, examiners do not usually appear to limit their consideration of a reference to its ability to support an anticipation rejection, to the neglect of its potential use in an obviousness rejection. There were only six cases (8 percent) in which the PTAB determined that a claim was unpatentable due to obviousness over a reference, used by the PTAB either alone or in combination with another reference, that the examiner had applied only in an anticipation rejection (question III.C). These findings are especially noteworthy because it is likely that the third party who filed the petition to begin the IPR proceeding would argue for unpatentability of the claims over such references. Although the data set is small, these results seem to suggest that once examiners recognize that a reference is relevant, they can competently apply it to the claims. This underscores the need to help examiners to locate and recognize appropriate prior art as discussed above.

6.2.4. The Examiner's Handling of Application Continuity

Section IV of the questionnaire relates to the proper treatment of continuity claims by the examiner. Fifty cases (64 percent) included a domestic benefit claim, 9 cases (12 percent) included a foreign priority claim, and 5 cases (6 percent) were U.S. national stage entries under 35 U.S.C. § 371 of international applications. Out of all 78 cases in the study group, a continuity claim was mishandled by the examiner in only one domestic benefit situation (Figure 6-7).

Despite the small data set, this result strongly suggests that examiner errors in handling continuity issues are not a significant driver of IPR filings.

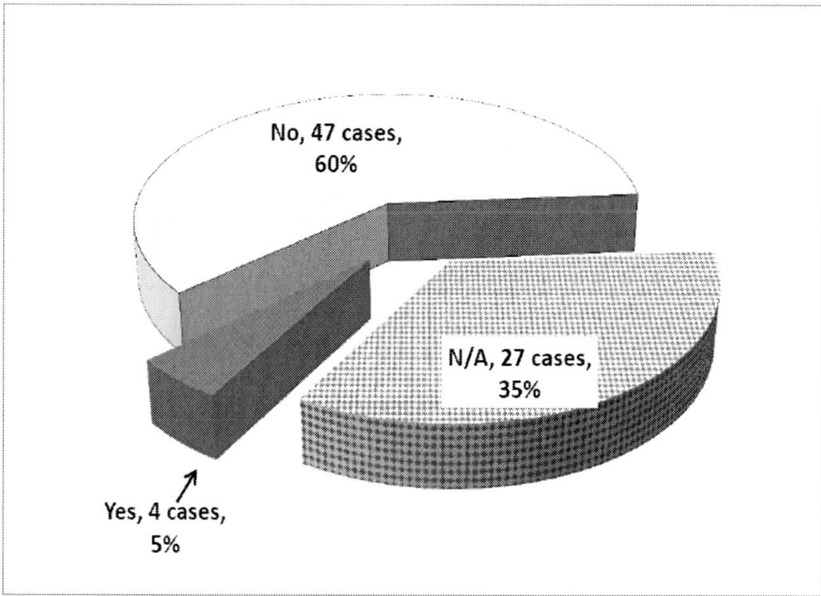

Apparent Reason the Examiner Withdrew the Rejection	102 Rejections		103 Rejections	
	Yes	No	Yes	No
Claim Interpretation	0	4	0	4
Consideration of applicant's arguments	4	0	1	3
Consideration of interview held	1	3	0	4
Other reason	0	4	0	4

Figure 6-6. Examiner's Withdrawn Art Later Relied on by PTAB.

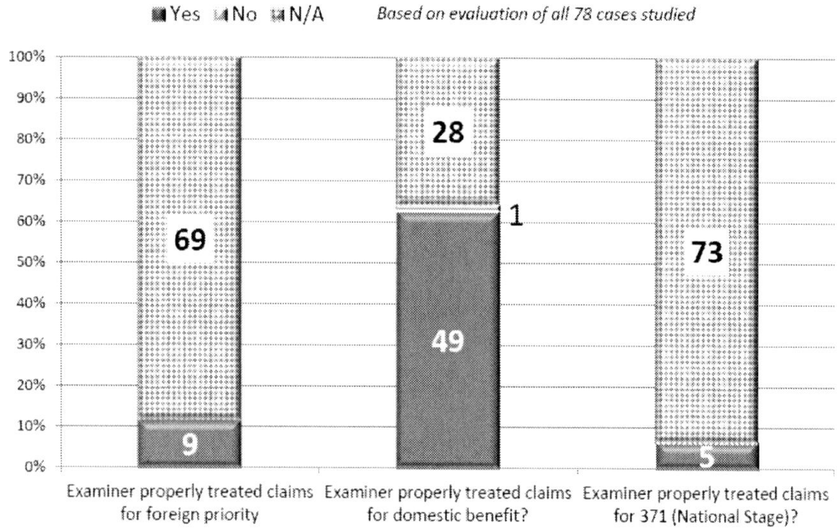

Figure 6-7. Examiner's Treatment of Continuity Claims.

6.2.5. Substantive Interviews between the Examiner and the Applicant

Section V of the questionnaire relates to the substantive interviews between the examiner and the applicant. The importance of examiner-applicant interviews to the patent prosecution process has been widely discussed, and the Office has published an "Interview Best Practices" document to promote productive interviews.[67] The Office continues to evaluate programs to enhance communication between examiners and applicants, such as the Pre-Appeal Brief Pilot Program[68] and the First Action Interview Pilot Program.[69] Thus, we hoped that the present hands-on review would provide useful information about the relationship between interviews and the incidence of IPR proceedings.

Regular patent prosecution procedure requires the examiner to ensure that a written record of any substantive interview be included in the prosecution history. However, the written record of an interview need not be indexed in PALM as a separate document, but may instead be included as part of another Office action. Therefore, in an effort to gather information about all substantive interviews in each prosecution history, we asked the reviewers to count the number of interviews that were separately indexed, as well as those that were not. Of the 78 cases, 3 cases had three or more interviews, 5 had two interviews, 9 had one interview, and 51 had no interviews at all. That is, in our data set, at least one interview formed part of the prosecution history in only

27 of 78 cases (35 percent). See Table 6-4 and Figure 6-8 for information about number and frequency distribution of interviews. Table 6- 5 provides data about our reviewers' assessment of the interviews.

Table 6-4. Number of Interviews and Recordation

	Number of Interviews	Number of Interviews as % of Total	Total Number of Interviews Separately Recorded in PALM
Examiner-initiated interviews	15	35.7	7
Applicant-initiated interviews	11	26.2	6
Unable to determine who initiated	16	38.1	7
Total	42	100.0	20

n=78 cases; 27 cases had at least 1 interview

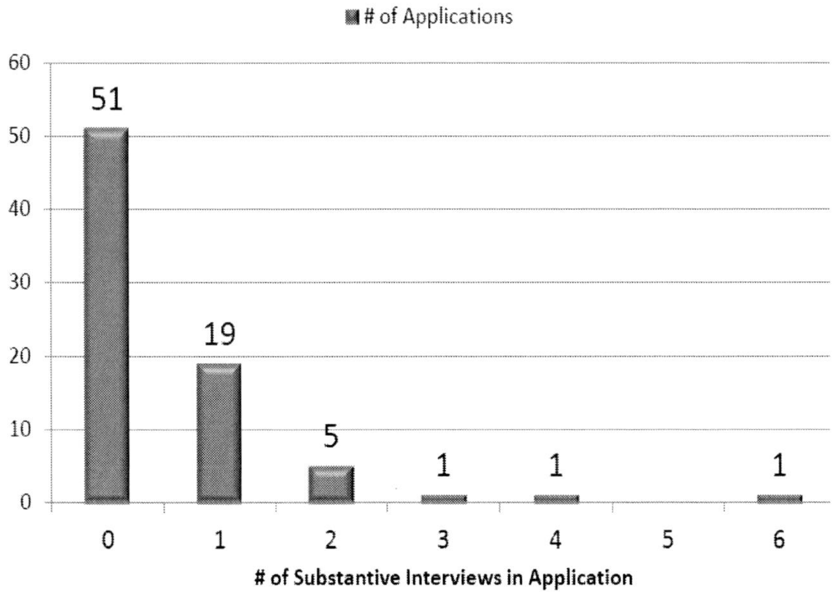

Figure 6-8. Frequency Distribution of Interviews.

In view of the particular attention that has been paid to interviews both by the Office and by our stakeholders, it would be interesting to determine whether this seemingly low percentage of cases which include at least one interview would also be observed if a larger sample size were studied. If so, it could be an indication of the positive value of interviews. It is possible that interviews lead to claims that are more clearly defined over the prior art, and that such claims are less likely to be challenged in an IPR proceeding.

Table 6-5. Reviewer Assessment of Examiner Interviews

	Yes		**No**		**N/A or Unable to Evaluate**	
	#	%	#	%	#	%
Examiner provided either a clear and complete summary of the interview or a statement of agreement with a summary provided by attorney	28	66.7	10	23.8	4	9.5
Interview resulted in an examiner's amendment	14	33.3	20	47.6	8	19.0
Interview indicated allowable subject matter	10	23.8	22	52.4	10	23.8
Interview provided correct suggestions to overcome rejection(s)	9	21.4	16	38.1	17	40.5
Applicant adopted exami ner suggestion resulting in allowance?	15	35.7	6	14.3	21	50.0

n=42 total interviews.

6.2.6. Claim Interpretation Issues

Section VI of the questionnaire relates to issues of claim interpretation. In 18 percent (14 cases) of the 78-member study group, the examiner and the PTAB disagreed on an issue of claim interpretation (Figure 6-9). Of these 14 cases in which there was a disagreement, the reviewers determined that the difference in claim interpretation had an impact on patentability in only 9 of them (12 percent of the total study group). Although this is a desirable result, once again the small sample size limits our ability to draw meaningful conclusions.

Furthermore, one could also observe that in 9 of the 14 cases (64 percent) in which there was a claim interpretation disagreement, that disagreement was crucial to the PTAB's patentability determination. In March 2014, the Office launched the Glossary Pilot Program, aimed at promoting claim clarity.[70] If successful, that pilot program could serve to reduce disagreements about claim construction by encouraging clear explanations of claim terminology.

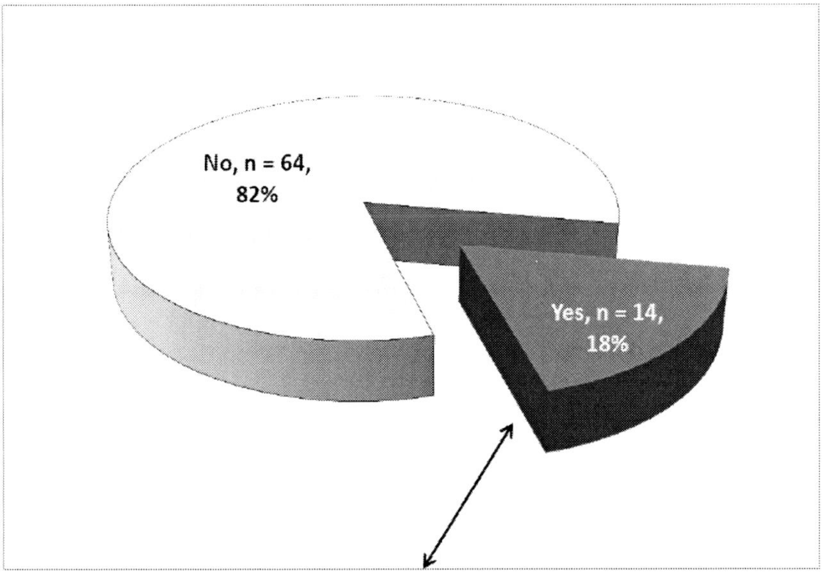

		Count	% of Total
If claim interpretation differed, did the difference have an impact on PTAB's conclusions regarding patentability?	Yes	9	64.3
	No	4	28.6
	Unable to Determine	1	7.1

Figure 6-9. Did Examiner's Claim Interpretation Differ from that of PTAB?

CONCLUSION

The purpose of this study was to respond to the recommendation of the GAO Report: namely, to link patent infringement litigation information to patent examination information, with the goal of discovering ways to enhance patent quality. We expanded the scope of the study to include IPR proceedings at the PTAB.

For all of the investigations described in this response to the GAO Report – the matched case-control studies of both litigated and IPR-petitioned patents, the comparison between instituted and non-instituted IPR-petitioned patents, and the hands-on review of the prosecution history of patents which had undergone IPR – any conclusions must be viewed in light of the changes that have occurred in patent law and practice over the last several years. For the infringement suits used in the matched case-control study, the issue dates of the patents involved ranged from 1988 to 2011. For the IPR proceedings used in the matched case-control study and the instituted vs. non-instituted comparison, the issue dates of the patents involved ranged from 1995 to 2013. For the hands-on review, the issue dates ranged from 1996 to 2012. Between 1988 and 2013, the United States Supreme Court weighed in on a number of patent cases that impact how a determination of patentability is made. The most notable are KSR v. Teleflex concerning the standard for obviousness, and Bilski v. Kappos, Association for Molecular Pathology v. Myriad Genetics, Inc., and Mayo Collaborative Services v. Prometheus Laboratories., Inc., concerning statutory subject matter.[71] In response, the USPTO issued guidance documents to ensure consistent practice by examiners across TCs, and to aid the public in understanding how the Office viewed the state of the law.[72] In the same time period, the United States Congress passed the American Inventor's Protection Act of 1999 (AIPA), the Cooperative Research and Technology Enhancement Act of 2004 (CREATE Act), and the Leahy-Smith America Invents Act of 2011 (AIA), and the USPTO issued corresponding implementing regulations.[73] The USPTO has also been active in instituting a variety of programs as temporary pilots or as ongoing additions to patent prosecution procedures. It is important to keep these developments in mind when interpreting the results of our studies.

Although patent infringement litigation in federal district court is a well-established procedure, IPR is a new proceeding created by the AIA. As we have already pointed out, the sample set for all of our IPR-related studies includes a rather small number of proceedings, over a time period beginning on September 16, 2012 when IPRs first became available, and ending in mid-July 2014.

Thus, the IPR proceedings that we have studied are part of a new endeavor, in which neither the PTAB, nor the patent holders, nor the third party petitioners have had any prior experience. Commentators have noted the high rate of unpatentability findings at the PTAB, and some have questioned whether the Board is overly aggressive.[74] At the same time, it is possible that the first wave of IPR cases reflects a type of "pent up demand." Thus, it is worthwhile monitoring whether the trends suggested by our data persist as the IPR proceeding matures.

Additionally, it is worth investigating whether any unpatentability decisions are the result of the examination at the time of grant, or the result of changes in standards between grant and petition.

The most compelling result from both the infringement litigation and IPR matched case-control studies is that the highest-impact variables are those over which the USPTO has little control, and which are not directly related to the examination process itself. As we explained in Sections 4.1.2 and 4.2.2, variables concerning entity size, foreign origin, and the number of domestic parent applications were the most important in predicting the likelihood of subsequent litigation and IPR proceedings. Models including only these variables perform almost as well as the models which include all of the variables that we considered. While the most impactful variables are defined prior to any examination, they can be indirectly affected by factors over which the Office has some control. For instance, small entity fees affect the incentives of small entities to seek patent protection. The level of the fees, then, impacts which applications are sought, and which patents are granted. Indirectly, this affects the rates of litigation that we later see. As another example, the examination history of a parent application could impact the characteristics of the child application, or whether a child application is even filed.

Continuing applications could also be correlated with higher patent value or with a greater propensity toward strategic behavior by applicants.[75] Also, it is interesting to note that unlike these continuing applications (i.e., applications with domestic parents), RCEs did not have any measureable impact on the likelihood of litigation or IPR petition.

The matched case-control studies also indicate that certain patent- and patent examination-related characteristics are correlated with the likelihood of subsequent patent infringement litigation and petitions for IPR. These results are particularly clear with regard to independent claims. Patents with more independent claims are more likely to be involved in subsequent litigation and IPR. At the same time, patents with fewer words per independent claim are also more likely to be involved in these post-grant challenges. More research is needed to determine whether the narrowness and clarity of scope is a function of the examination quality or whether it is inherent to the specific invention for which patent protection is sought. For instance, inventions in new or evolving technology areas may involve more intrinsic uncertainty, because they may not have well-developed lexicons and it may be difficult for the courts to apply previous case law to new cases within that field.

Other patent examination-related characteristics that our matched case-control studies suggest may be relevant include the GS-level of the allowing examiner, the number of IDS filings, the number of examiner interviews, and whether the patent was allowed without receiving a previous rejection. Again, while statistically significant, the impact of these variables is much smaller in magnitude than application characteristics like small entity status and foreign priority.

We find that patents allowed by lower GS-level examiners are generally less likely to be either asserted by the patent owner in an infringement suit or petitioned for IPR by someone other than the patent owner. The same can be said of patents allowed on first action, at least with regard to litigation (although not with regard to IPRs). IDS filings tend to be positively associated with a greater chance of litigation and IPRs, while interviews tend to be positively associated with litigation. It is possible that applicants will be more likely to file IDSs and to conduct interviews with examiners when trying to secure allowance of applications that they believe are more valuable. In other words, these two variables may signal a higher value of the patent rights involved. On the other hand, applications that are allowed on the first action are likely to contain clearer and more narrowly drafted claims. These factors are likely to reduce the number of challenges to patentability and to cause fewer disputes about claim scope. We hope that further research can delve more deeply into these matters.

In our comparison between instituted and non-instituted IPR-petitioned patents, there were very few statistically significant results. This indicates that the observable characteristics that we identified do not do much to explain the difference between instituted and non-instituted patents. The implication is that the unobserved differences in the individual patents explain more of the variation between institution and non-institution than the observed and measured differences. The most significant variable was the GS-level of the examiner at the time the case was allowed. However, even here there is some ambiguity. We found that if the examiner who allowed the claims was GS-13, the patent was less likely to have had the IPR proceeding instituted than when the examiner was GS-14, *or GS-12 or below.*[76] Another result that emerged from our comparison study was that patents issued in 2000-2004 are more likely to have an IPR proceeding instituted than patents issued since then.

The most notable results from the hands-on study concerned the sources of prior art used by the PTAB to find a claim unpatentable, and the examiner's prior art search.

The prior art relied upon by PTAB to hold a claim unpatentable had been available during prosecution in 59 percent of the cases where PTAB applied art. It may be worthwhile to monitor the decisions to determine the extent to which changes in law explain the difference in interpretation between the examiner and the Board. Our data, while not conclusive, suggest that the number of references cited on the IDS may have been a factor; that is, citation of a large number of references may impede the examiner's ability to identify those that are most relevant to the claims under examination. When PTAB determined that a claim was unpatentable in view of prior art not located by the examiner's search, OPQA reviewers considered the examiner's search to have been deficient in 61 percent of the cases. As we explain in Section 6.2.2, the self-selection of cases with sub-par searches for submission of IPR petitions, as well as possible hindsight bias on the part of OPQA reviewers, cause us to question the generalizability of this result to the average patent examination.

It is interesting to consider the results of our matched case-control study of the likelihood of patent litigation in light of the results obtained by Cockburn et al. (2003). As we have described above, our data set of litigated patents included those for which an infringement suit was filed in district court, regardless of the eventual disposition of the case. Cockburn et al., on the other hand, looked at instances of patent litigation in which there had been a ruling on validity by the CAFC. In a sense, ours is an *a priori* data set, while that of Cockburn et al. is *a posteriori*. Both our study and that of Cockburn et al. (2003) included an assessment of whether the experience level of the allowing examiner had an impact on whether a case would eventually become a member of the respective pools of cases under study. Although our results generally indicated that patents allowed by more experienced examiners were more likely to be the subject of a patent infringement lawsuit, the Cockburn et al. results did not find a correlation between the experience level of the allowing examiner and the likelihood that the CAFC would find the patent to be invalid.[77]

This comparison raises more questions, and suggests further research. Because the Cockburn et al. study was conducted more than ten years ago, it is possible that the impact of examiner experience may have changed over time. Otherwise, it would lead to the question of why more experienced examiners are more likely to have their patents litigated, but less likely to have them invalidated. Further research along these lines could provide helpful insights for enhancing patent quality.

Lastly, it is important to remember that most patent infringement suits settle, and only a minority of claims proceeds to adjudication. Further, not all adjudications include a ruling on validity. As explained above, for this preliminary study, it was expedient to focus on the incidence of a filed complaint, because of the availability of data. Because invalidity of the asserted patent is a common defense against infringement, it is not unreasonable to think that our dataset encompassed patents that were subject to a validity challenge. The mere filing of an infringement suit certainly implies a failure of the parties of a dispute to come to agreement without resorting to formal dispute resolution. However, lawsuits that are dismissed or that quickly settle are clearly different from those that proceed all the way through trial.

More research is necessary to investigate the multiple layers of selection that filter the cases throughout the litigation process, and the relationship to patent examination characteristics. Thus, one should be cautious in generalizing the results of this study to all patents.

This study provided some preliminary findings tying patent examination characteristics to patent infringement litigation and inter partes review petitions and proceedings. While some results are provocative, more research will be necessary in order to fully understand the implications for patent quality. Because the variables outside the direct control of the examiner (entity size, foreign origin, and continuation history) are most important in predicting the incidence of subsequent litigation and IPR petitions, it does not immediately suggest any specific avenues for the USPTO to explore with regard to enhancing patent quality.

However, efforts aimed at improving the overall examination process in order to reduce the need for continuing applications could ultimately lead to fewer litigated patents. The results of the IPR hands-on study underscore the need for examiners to be able to identify references that are relevant to patentability. The USPTO will continue to work in partnership with applicants and practitioners to ensure that examiners utilize the best prior art available during examination. We expect that continued research along these lines will result in useful initiatives for enhancing patent quality.

8. GLOSSARY

AIC	Akaike information criterion
BIC	Bayesian information criterion

BIO	Biotechnology and organic chemistry
BPAI	Board of Patent Appeals and Interferences (predecessor of PTAB)
CHEM	Chemical and materials engineering
COMP	Computers and communications
CON	Traditional continuation application (serialized continuation)
CPA	Continued Prosecution Application
DES	Designs
DIV	Divisional application
EPO	European Patent Office
FAOM	First action on the merits
GAO	United States Government Accountability Office
GAU	Group art unit
GS	General Schedule (GS) classification and pay system (Office of Personnel Management)
HR	Human resources
IDS	Information disclosure statement
IPC	International Patent Classification
IPR	Inter partes review
MECH	Mechanical engineering, manufacturing, products
NBER	National Bureau of Economic Research
OPQA	Office of Patent Quality Administration
PALM	Patent Application Location and Monitoring system
PTAB	Patent Trial and Appeal Board (successor to BPAI)
RCE	Request for continued examination
Report	"Intellectual Property: Assessing Factors that Affect Patent Infringement Litigation Could Help Improve Patent Quality"
SEMI	Semiconductors, electrical and optical systems and components
STIC	Scientific and Technical Information Center
TAF	Technology Assessment and Forecast
TC	Technology center
TRANS	Transportation, construction, electronic commerce, agriculture, national security and license & review
USPTO or Office	United States Patent and Trademark Office

9. Appendix I. IPR Proceedings Studied in Hands-on Review

PTAB Trial #	Patent #	Patent Issue Date	TC	Patent Owner	Number of Claims in Patent	Number of Claims found Unpatentable
IPR2012-00001	6778074	8/17/2004	2600	Cuozzo Speed Technologies LLC	20	3
IPR2012-00005	6653215	11/25/2003	2800	Emcore Corporation	17	17
IPR2012-00006	7713698	5/11/2010	1600	National Science Foundation	17	12
IPR2012-00007	7790869	9/7/2010	1600	National Science Foundation	33	16
IPR2012-00018	7566960	7/28/2009	2800	Xilinx, Inc.	13	13
IPR2012-00019	8062968	11/22/2011	2800	Xilinx, Inc.	15	15
IPR2012-00020	8058897	11/15/2011	2800	Xilinx, Inc.	14	11
IPR2012-00023	7994609	8/9/2011	2800	Xilinx, Inc.	19	5
IPR2012-00027	7591303	9/22/2009	3700	Bergstrom, Inc.	23	14
IPR2012-00042	6240376	5/29/2001	2100	Mentor Graphics	33	3
IPR2013-00005	6444533	9/3/2002	2800	The Board of Trustees of the University of Illinois	8	8
IPR2013-00006	6888204	5/3/2005	2800	The Board of Trustees of the University of Illinois	18	18
IPR2013-00008	5872387	2/16/1999	2800	Dept. of Navy	2	2
IPR2013-00010	7516484	4/7/2009	2400	Michael Arnouse	20	7
IPR2013-00011	8088575	1/3/2012	1600	National Science Foundation	6	4
IPR2013-00014	6998973	2/14/2006	2600	Continental Automotive Systems	12	10
IPR2013-00016	6441828	8/27/2002	2600	Mobile Media Ideas	18	0

PTAB Trial #	Patent #	Patent Issue Date	TC	Patent Owner	Number of Claims in Patent	Number of Claims found Unpatentable
IPR2013-00020	7297364	11/20/2007	1700	Clearlamp	24	12
IPR2013-00026	6029111	2/22/2000	3600	Beacon Navigation	28	11
IPR2013-00029	5632545	5/27/1997	2800	Intellectual Ventures	3	3
IPR2013-00033	7155241	12/26/2006	2600	Helferich Patent Licensing	94	41
IPR2013-00034	7970674	6/28/2011	3600	Silicon Valley Bank	40	25
IPR2013-00041	5746775	5/5/1998	1700	NIH	32	6
IPR2013-00043	7171103	1/30/2007	2800	DSM IP Assets B.V.	18	0
IPR2013-00044	6961508	11/1/2005	2800	DSM IP Assets B.V.	22	0
IPR2013-00045	6339666	1/15/2002	2800	DSM IP Assets B.V.	20	0
IPR2013-00046	6110593	8/29/2000	1700	DSM IP Assets B.V.	9	9
IPR2013-00047	6438206	10/23/2002	2800	DSM IP Assets B.V.	14	0
IPR2013-00048	6298189	10/2/2001	2800	DSM IP Assets B.V.	66	8
IPR2013-00050	6323255	11/27/2001	1700	DSM IP Assets B.V.	19	19
IPR2013-00053	7276543	10/2/2007	1700	DSM IP Assets B.V.	57	8
IPR2013-00057	6074503	6/13/2000	1700	Belden Technologies, Inc.	6	4
IPR2013-00058	7977575	7/12/2011	2800	Belden Technologies, Inc.	34	34
IPR2013-00059	7135641	11/14/2006	2800	Belden Technologies, Inc.	22	20
IPR2013-00063	6513058	1/28/2003	2100	Roy-G-Biv Corp.	5	0
IPR2013-00067	7347011	3/25/2008	3700	Nike, Inc.	46	0
IPR2013-00069	7663061	2/16/2010	2800	Belden Technologies, Inc.	21	21

Appendix I. (Continued)

PTAB Trial #	Patent #	Patent Issue Date	TC	Patent Owner	Number of Claims in Patent	Number of Claims found Unpatentable
IPR2013-00072	D617465	6/8/2010	2900	Luv N' Care, Ltd.	1	1
IPR2013-00080	6173403	1/9/2001	2400	Achates	19	15
IPR2013-00081	5982889	11/9/1999	2400	Achates	4	4
IPR2013-00082	5978791	11/2/1999	2100	PersonalWeb Tech.	48	10
IPR2013-00083	6415280	7/2/2002	2100	PersonalWeb Tech.	55	2
IPR2013-00084	7945544	5/17/2011	2100	PersonalWeb Tech.	56	1
IPR2013-00085	7945539	5/17/2011	2100	PersonalWeb Tech.	34	3
IPR2013-00086	7949662	5/24/2011	2100	PersonalWeb Tech.	35	1
IPR2013-00087	8001096	8/16/2011	2100	PersonalWeb Tech.	137	4
IPR2013-00093	6065880	5/23/2000	2800	Mobile Scanning Tech.	20	3
IPR2013-00097	6669981	12/30/2003	1700	Convatec Inc.	20	19
IPR2013-00102	7267828	9/11/2007	1600	Convatec Inc.	14	13
IPR2013-00106	6526808	3/4/2003	2800	Star EnviroTech	10	0
IPR2013-00109	6757717	6/29/2004	2100	Proxyconn Inc.	34	6
IPR2013-00110	6209259	4/3/2001	1600	Encap	14	12
IPR2013-00111	6494387	12/17/2002	3700	KANEKO	22	14
IPR2013-00112	5779334	7/14/1998	2800	Intellectual Ventures	14	10
IPR2013-00116	5997915	12/7/1999	1700	Merck	121	8

PTAB Trial #	Patent #	Patent Issue Date	TC	Patent Owner	Number of Claims in Patent	Number of Claims found Unpatentable
IPR2013-00117	6011040	1/4/2000	1600	Merck	22	10
IPR2013-00118	6673381	1/6/2004	1700	Merck	39	1
IPR2013-00119	7172778	2/6/2007	1700	Merck	15	1
IPR2013-00121	7731558	6/8/2010	3700	Capriola	27	8
IPR2013-00124	7579016	8/25/2009	1600	US Dept. of Agriculture	26	1
IPR2013-00127	6587067	7/1/2003	2600	Universal Electronics	5	6
IPR2013-00133	7523072	4/21/2009	3600	ContentGuard Holdings	25	0
IPR2013-00137	6963859	11/8/2005	3600	ContentGuard Holdings	34	0
IPR2013-00138	7139736	11/21/2006	3600	ContentGuard Holdings	57	0
IPR2013-00139	7269576	9/11/2007	3600	ContentGuard Holdings	36	14
IPR2013-00155	8180489	5/15/2012	1700	Culligan International Co.	6	6
IPR2013-00170	6581065	6/17/2003	2100	GE	42	42
IPR2013-00175	7633673	12/15/2009	3600	Cheetah Omni	20	7
IPR2013-00179	7712683	5/11/2010	1700	Prolitec	2	2
IPR2013-00196	8013884	9/6/2011	2800	Automatic Manuf. Systems	20	20
IPR2013-00208	8251997	8/28/2012	3700	Warsaw Orthopedic	30	8
IPR2013-00209	8317070	11/27/2012	3700	Ethicon Endo-Surgery	14	14
IPR2013-00252	5495291	2/27/1996	2600	Avid Technology	20	10
IPR2013-00256	7461353	12/2/2008	2100	Softview	319	18
IPR2013-00257	7831926	11/9/2010	2100	Softview	88	10

Appendix I. (Continued)

PTAB Trial #	Patent #	Patent Issue Date	TC	Patent Owner	Number of Claims in Patent	Number of Claims found Unpatentable
IPR2013-00282	6516236	2/4/2003	2100	Roy-G-Biv Corp.	10	0
IPR2013-00286	8073557	12/6/2011	2100	Roy-G-Biv Corp.	59	0
IPR2013-00495	6218930	4/17/2001	2600	Network-Security Solution, Inc.	9	0

10. Appendix II. Questionnaire for Hands-on Review by Quality Assurance Specialists at USPTO

PTAB Inter Partes Review: Final Written Decisions Root Cause Analysis

PTAB Trial #: Patent #: Application #:

I. Prior Art Found

A. PTAB determined that some or all of the patented claims are unpatentable over (check all that apply):

n=64 cases where at least 1 claim found unpatentable in FWD.

45%	Art previously of record, but not used as basis for rejection during examination
20%	Art previously of record that was used as basis for rejection that was withdrawn by examiner during examination
95%	New art identified during Inter Partes Review

B. Prior art identified during Inter Partes Review and determined by PTAB to render claims unpatentable (check all that apply):

n=61 cases where at least 1 claim found unpatentable in FWD and new art was identified in Inter Partes Review

84%	US Patent or application publication
43%	Foreign patent document
33%	NPL

Appendix II. (Continued)

C. Characteristics of art determined to render claims unpatentable in Inter Partes Review vs. any art applied by examiner during prosecution (check all that apply):

n=64 cases where at least 1 claim found unpatentable in FWD.

3%	has a better date (e.g. 102(b) versus 102(a) or (e)).
9%	is a 102 whereas the examiner applied only 103 art
12%	covers more elements of the claims such that fewer references are needed in the combination
8%	provides a better reason to combine
50%	shows elements of the claim that are not provided for in the art applied by the examiner
17%	requires a greater narrowing of the claim(s) to overcome rejection
31%	no art applied during prosecution

Comments:

II. Search

A. If prior art determined to render the claims unpatentable was not uncovered by examiner's search, was the search performed considered to be deficient? 61% Yes 36% No 3% N/A

n=61 cases where at least 1 claim found unpatentable in FWD and new art was identified in Inter Partes Review

B. If the examiner's search was considered deficient, check any that apply:

68%	Search is too limiting	24%	Poor use of proximity and Boolean operators
22%	Required class/subclass was not included	46%	No inventor's name search
		49%	Other

Comments:

n=37 cases where search by examiner was considered deficient

PTAB Inter Partes Review: Final Written Decisions Root Cause Analysis

II. Search *continued*

C. Did the examiner do an appropriate search in each of the following areas? [*n=78; all cases with FWD.*]

	Yes	No	N/A
US Patents and PGPubs	77%	23%	
Foreign patent documents	49%	51%	
Non-Patent Literature (e.g. Dialog, STN, IEEE, Google, etc.)	29%	71%	
Inventor name search performed	29%	71%	
Specialized tools relevant to the technology (nucleic acid sequence search, chemical structure search)	*Insufficient data points for summary*		

D. Did the examiner properly consider all IDS submitted during prosecution? [*n=78; all cases with FWD.*]

68% Yes 23% No 9% N/A

E. Was prior art determined by PTAB to render claims unpatentable cited in an IDS during prosecution?

87% Yes 13% No 0% N/A [*n=38where at least 1 claim found unpatentable in FWD and art previously of record*]

III. Prior Art Rejections

A. Were there any 102 or 103 rejections made by the examiner and subsequently withdrawn that the PTAB decision indicated should have been maintained? 5% Yes 61% No 35% N/A [*n=78; all cases with FWD.*]

Appendix II. (Continued)

B. If yes, check any that apply as to the apparent reason that the examiner withdrew the rejection.

n=4; there was a 102 or 103 rejection made by examiner and subsequently withdrawn but PTAB decision indicated it should have been maintained

	102 Rejection	103 Rejection
Claim interpretation	**0%**	**0%**
Consideration of applicant's arguments	**100%**	**25%**
Consideration of interview held	**25%**	**0%**
Other	**0%**	**0%**

Provide a brief explanation for any box checked:

C. Did the examiner apply a reference only in a 102 rejection which the PTAB determined should have been used in a 103 rejection? 8% Yes 50% No 42% N/A

n=78; all cases with FWD. N/A=no 102 rejection made in case

IV. Continuity

n=78; all cases with FWD.

	Yes	No	N/A
A. Examiner properly treated claims for foreign priority?	12%	0%	88%
B. Examiner properly treated claims for domestic benefit?	63%	1%	36%
C. Examiner properly treated claims for 371 (National Stage)?	6%	0%	94%

If no to any of the above, please provide a brief explanation, including an indication of whether the improper treatment affected the outcome of the PTAB:

PTAB Inter Partes Review: Final Written Decisions Root Cause Analysis

V. Substantive Interviews

n=78; all cases with FWD.

A.	# of substantive Examiner-initiated interviews	15	# that were separately recorded in PALM	7
B.	# of substantive Applicant-initiated interviews	11	# that were separately recorded in PALM	6
C.	# of substantive interviews unable to determine who initiated	16	# that were separately recorded in PALM	7

51 cases with no interview; 19 cases with 1 interview; 5 cases with 2 interviews; 3 cases with 3 or more interviews

D.	# Yes	# No	# N/A or unable to evaluate
i. Examiner provided either a clear and complete summary of the interview or a statement of agreement with a summary provided by attorney	28	10	4
ii. Interview resulted in an examiner's amendment	14	20	8
iii. Interview indicated allowable subject matter	10	22	10
iv. Interview provided correct suggestions to overcome rejection(s)	9	16	17
v. Applicant adopted examiner suggestion resulting in allowance?	15	6	21

E. Were issues raised during Inter Partes Review identified in any interview during examination?

3%	Yes	28%	No	4%	Unable to Determine	65%	N/A

n=78; all cases with FWD.
N/A=no interview in 51 cases

Comments:

Appendix II. (Continued)

VI. Claim Interpretation

A. Did the examiner's claim interpretation differ from the PTAB's? 18% Yes 82% No [*n=78; all cases with FWD.*]

B. If yes, did the difference have an impact on PTAB's conclusions regarding patentability?

64% Yes 29% No 7% Unable to Determine [*n=14 where claim interpretation differed between PTAB and examiner*]

Please provide comments about the reasonableness of claim interpretation by the examiner and what, if any, effect it may have had on the PTAB outcome.

VII. Root Cause

Please provide a succinct summary of the reasons that the examiner did not reach the same conclusions as PTAB regarding claims that were found unpatentable.

REFERENCES

Allison, John R., Mark A. Lemley, Kimberly A. Moore, and R. Derek Trunkey. 2004. "Valuable Patents." *Georgetown Law Journal* 92:435.

Allison, John R., Mark A. Lemley, and Joshua Walker. 2009. "Extreme Value or Trolls on Top? The Characteristics of the Most-Litigated Patents." *University of Pennsylvania Law Review* 158: 1.

Allison, John R., Emerson H. Tiller, and Smantha Zyontz. 2012. "Patent Litigation and the Internet." *Stanford Technology Law Review* 2012: 1.

Chien, Colleen V. 2011. "Predicting Patent Litigation." *Texas Law Review* 90: 283.

Cockburn, Iain M., Samuel Kortum, and Scott Stern. 2003. "Are All Patent Examiners Equal? Examiners, Patent Characteristics, and Litigation Outcomes." In *Patents in the Knowledge-Based Economy*, edited by Wesley Cohen and Stephen Merrill, Washington, DC: National Academies Press.

Cooter, Robert D. and Daniel L. Rubinfeld. 1989. "Economic Analysis of Legal Disputes and Their Resolution." *Journal of Economic Literature* 27 (3): 1067-1097.

Cox, D. R. and D. V. Hinkley. 1974. *Theoretical Statistics*. 1st ed. London, UK: Chapman and Hall.

Government Accountability Office. 2013. "Intellectual Property: Assessing Factors That Affect Patent Infringement Litigation Could Help Improve Patent Quality." GAO Report GAO-13-465.

Hall, Bronwyn, Adam Jaffe, and Manuel Trajtenberg. 2001. "The NBER Patent Citations Data File: Lessons, Insights, and Methodological Tools." National Bureau of Economic Research (NBER) Working Paper 8498, October.

Hosmer, D.W., S. A. Lemeshow, and R. X. Sturdivant, 2013. *Applied Logistic Regression*. 3rd ed. Hoboken, NJ: Wiley.

Kesan, Jay P., David L. Schwartz, and Ted Sichelman. 2012. Paving the Path to Accurately Predicting Legal Outcomes: A Comment on Professor Chien's Predicting Patent Litigation. *Texas Law Review* 90: 97.

King, John L. 2003. "Patent Examination Procedures and Patent Quality." In *Patents in the Knowledge-Based Economy*, edited by Wesley Cohen and Stephen Merrill, Washington, DC: National Academies Press.

Kupper L.L., J.M. Karon, D.G. Kleinbaum, H. Morgenstern, and D.K. Lewis. 1981. "Matching in Epidemiologic Studies: Validity and Efficiency Considerations." *Biometrics* (37): 271–291.

Lanjouw, Jean O. and Mark Schankerman. 2001. "Characteristics of Patent Litigation: A Window on Competition." *RAND Journal of Economics* 32 (1): 129-151.

Lanjouw, Jean O. and Mark Schankerman. 2004. "Protecting Intellectual Property Rights: Are Small Firms Handicapped?" *Journal of Law and Economics* 47 (1): 45-74.

Lerner, Josh. 2008. "The Litigation of Financial Innovations." NBER Working Paper 14324.

Love, Brian J. and Ambwani, Shawn. 2014. "Inter Partes Review: An Early Look at the Numbers." *University of Chicago Law Review Dialogue* 81: 93.

Maddala, G. S. 1983. *Limited Dependent and Qualitative Variables in Econometrics*. Cambridge, U.K: Cambridge University Press.

Miller, Shawn P. 2013. "What's the Connection Between Repeat Litigation and Patent Quality? A (Partial) Defense of the Most Litigated Patents." *Stanford Technology Law Review* 1: 313.

Mitra-Kahn, Benjamin, Alan Marco, Michael Carley, Paul D'Agostino, Peter Evans, Carl Frey, and Nadiya Sultan. 2013 "Patent backlogs, inventories and pendency: An international framework," IPO & USPTO joint report. ISBN 978-1-908908-70-4. https://www.gov.uk/government/uploads/system/uploads/attachment data/file/311239/ipresearch -uspatlog-201306.pdf.

Patent and Trademark Office. 2014. *Manual of Patent Examining Procedure*. 9th ed. United States Department of Commerce. http://www.uspto.gov/web /offices/pac/mpep/index.html (accessed December 1, 2014).

Petherbridge, Lee. 2012. "On Predicting Patent Litigation." *Texas Law Review* 90: 75.

Rosenbaum, P. and D. B. Rubin. 1983. "The Central Role of the Propensity Score in Observational Studies for Causal Effects," *Biometrika* 70(1): 41-55.

Rothman, K. and S. Greenland. 1998. *Modern Epidemiology*. 2nd Edition, Philadelphia, PA: Lippincott, Williams and Wilkins.

Tu, Sean. 2012. "Luck/Unluck of the Draw: An Empirical Study of Examiner Allowance Rates." *Stanford Technology Law Review* 2012: 10.

Tu, Shine. 2014. "Patent Examiners and Litigation Outcomes." *Stanford Technology Law Review* 17: 507.

Taylor, Robert Brendan. 2012. "Burying." *19 Mich. Telecomm. & Tech. L. Rev. 99.*

End Notes

[1] See Cooter and Rubinfeld (1989) for an early survey.

[2] A patent's forward citations are citations to that patent by future patents. A patent's backward citations are citations in that patent to previous patents. The number of forward citations is often used as a measure of patent value.

[3] Examination groups are the precursors of the current technology centers.

[4] Lerner uses the term "financial patent" to refer to a patent that is classified in any of Class 705 subclasses 4 or 35 through 45 or Class 902 subclasses 1 through 41.

[5] See the Manual of Patent Examining Procedure (MPEP) 601.01. A filing date is assigned when the application is complete.

[6] Technology centers are comprised of work groups which are further comprised of group art units.

[7] If the applicant wishes to pursue patent protection on the additional inventions that are not chosen, one or more divisional applications may be filed. Such divisional applications retain the benefit of the filing date of the original application, and therefore have a longer pendency from filing of the original application to issue, even though the prosecution of the divisional application itself may not have been particularly lengthy.

[8] If the examiner decides to allow all claims at this stage, the action taken by the examiner is referred to as a first-action allowance.

[9] See Mitra-Kahn et al. (2013).

[10] This is not an applicant's only opportunity to abandon an application as applications may be abandoned at any time.

[11] CONs may be pursued at any time during prosecution. CONs may not contain new matter. New matter may be included in continuations-in-part (CIPs).

[12] Prior to the introduction of the RCE, applicants could file Continued Prosecution Applications (CPAs). In both cases, the continuations maintain the same serial number and the same specification as the original application.

[13] Prior to the AIA, the PTAB was known as the Board of Patent Appeals and Interferences, or BPAI.

[14] See, for instance Rothman and Greenland (1998).

[15] We chose these data for the following reasons. First, they constitute a random sample of recent patent infringement suits. Second, the data had already been vetted by the analysts at the GAO. Finally, such data are very valuable assets to the companies that collect them and obtaining full lists of litigation cases can be extremely costly. Lex Machina gave GAO permission to share these data with us for the purposes of this study. For future research, we intend to gather information on the entire population of litigated patents.

[16] Lex Machina maintains a database of patent infringement lawsuits filed in U.S. federal district courts since the year 2000, https://lexmachina.com/. The information is obtained from electronic court filings with these courts.

[17] The examiner's docket is the list of applications that are available for the examiner to work on.

[18] For the purpose of this study, we include continued prosecution applications (CPA) together with RCEs. Although a CPA is technically a new application while an RCE is not, both are ways for a patent applicant to continue to seek patent protection before the examiner for a particular invention, while retaining the same application serial number, and both are more similar to each other and different from CONs.

[19] See http://www.uspto.gov/web/offices/ac/ido/oeip/taf/brochure.htm#OEIP-Data_Elements.

[20] The data can be found at http://patents.reedtech.com/pgrbft.php.

[21] Note that the examiner is the one to whom the application was docketed at the time that the Notice of Allowance was mailed. This may be an examiner who does not have the independent authority to allow claims, and whose work is therefore overseen by a supervisor or primary examiner with signatory authority.

[22] In order for patent protection to be maintained throughout the full patent term, regular maintenance fee payments must be made to USPTO. This allows owners of less valuable patents to voluntarily cede their protection in exchange for lower lifetime fees. The maintenance fee payments are due 3.5, 7.5, and 11.5 years after the patent issues. In order to control for patent value and to make certain that the patent would still have been in force and thus could have been litigated, our matching algorithm matches on the number of maintenance fee payments made after matching on year of issue.

[23] We chose to generate two separate control groups as a way of checking of the robustness of our results.

[24] We choose the 3-year forward citation count because, although we match on year of grant, we do not have information on when each patent in the GAO sample was first litigated. Most were litigated more than three years from date of issuance.

[25] See Rosenbaum and Rubin (1983) for a discussion of propensity score matching.

[26] Interaction terms are created by multiplying the indicator variables for year of issue by the 3-year forward citation count. This allows the citation count to have a different impact for each grant year.

[27] As these constitute two of the exact matching variables, the issue year-technology mix of the patents in the control group is identical.

[28] In prior years, TCs 1200 and 1800 (both no longer in use) mapped to the BIO area,

[29] In prior years, TCs 1100, 1300, and 1500 (all no longer in use) mapped to the CHEM area.

[30] In prior years, TCs 2300 and 2700 (both no longer in use) mapped to the COMP area.

[31] In prior years, TCs 2200 and 2500 (both no longer in use) mapped to the SEMI area. Prior to 1998, TC 2100 mapped to the SEMI area instead of the COMP area.

[32] In prior years, TCs 3100 and 3500 (both no longer in use) mapped to the TRANS area.

[33] In prior years, TCs 3200, 3300, and 3400 (all no longer in use) mapped to the MECH area.

[34] See the section on the description of the GAO sample for the definitions of the technology areas.

[35] These outliers do not have a significant impact on our results when using the propensity score-matched sample.

[36] An indicator variable takes on different values based on whether an event occurred. Typically, it will be equal to one if the event occurred and zero if the event did not occur.

[37] We expect these variables to be positively correlated, but are interested in seeing which does a better job at predicting litigation.

[38] Sometimes examiner interviews are not separately indexed in PALM because they are appended to other documents. Thus, our reliance on PALM data undercounts the number of examiner interviews.

[39] It is important to note that here we are measuring the examination pendency of the application that was ultimately issued as a patent. The application may have been a continuation of a previous application. Any examination pendency inherent in the parent applications would be captured in the measure of pendency from the application date of the earliest parent application to the docketing date of the issued application.

[40] Claims are usually amended with each round of repeated negotiation.

[41] Currently the fee is $1200 for the first RCE in an application, and $1700 for the second and subsequent RCEs.

[42] Signatory authority is granted in stages, and includes two evaluation periods called the partial signatory authority program and the full signatory authority program. A newly-promoted GS-13 examiner, before beginning the partial signatory program, has no signatory authority. When the examiner begins the partial signatory authority program, she is granted temporary partial signatory authority. Upon successful completion, the partial signatory authority becomes permanent. When the examiner begins the full signatory authority program, she is granted temporary full signatory authority. Upon successful completion, the examiner is promoted to GS-14, and full signatory authority becomes permanent. However, a GS-13 examiner with permanent partial signatory authority may also be promoted to GS-14 by establishing Master's level competence in the technology of her assigned docket. Thus, not all GS-14 examiners have full signatory authority.

[43] Lanjouw and Schankerman use the total number of claims. Because dependent claims do not broaden the scope of their independent claims, we believe that the number of independent claims is more precise.

[44] Applicants are charged fees for independent claims in excess of three (currently $420 per claim).

[45] The estimation of logistic models relies on maximum likelihood estimation; and, the asymptotic optimality of maximum likelihood holds only when the number of parameters remains fixed. This is certainly not the case in any matched study. For matched studies, as the number of observations increases, the number of parameters also increases at the same rate. Thus, employing maximum likelihood to estimate the fully-stratified logistic model will lead to inconsistent results (Hosmer et al., 2013). However, if we are willing to forgo estimating the 974 matched-pair parameters, we can use conditional maximum likelihood to estimate the logistic model. Cox and Hinkley (1974) showed that such a method will yield consistent estimates of the coefficients when using 1-to-1 matched samples.

[46] A 1-percent level of statistical significance indicates that there is a less-than-1-percent chance that the two means are actually equal to one-another.

[47] We use these variables in the propensity score model because the literature suggests that they would be good predictors of litigation, and thus provide a good match, while not being directly related to the examination process.

[48] An odds ratio equal to one would indicate no relationship between the explanatory variable and the likelihood of litigation. A general rule of thumb regarding t-statistics is that an absolute value of more than 1.64 indicates marginal significance, an absolute value of greater than 1.96 indicates significance, and an absolute value of more than 2.58 indicates high significance.

[49] The statistically significant impacts are those with the markers filled in.

[50] For instance, the size of the small entity discount can influence the propensity of small entities to seek patent protection. As another example, the examination processes for progenitor applications may impact the incentives of applicants to file continuing applications. Thus, the examination processes of the two applications are not independent.

[51] The BIC for Model 1 is equal to 1133.5, whereas the BIC for the model constrained to just the entity size, foreign origin and continuation history is equal to 1077. The model with the lower value for the BIC is preferred. The corresponding AICs are 1010.8 and 1049.1. Again, the model with the lower AIC is preferred.

[52] Logistic models use a maximum likelihood approach to generate their parameter estimates. When choosing among closely related models, one can compare the estimated log likelihoods of the models (which are negative) and choose the largest one (or the smallest in absolute value). The AIC is a function of the estimated log likelihood and the number of

parameters (k) included in the model, namely -2*(log likelihood) + 2*k. In this case the best model is the one with the lowest AIC (since we are now multiplying the log likelihood by a negative number). The AIC penalizes a model for using additional parameters to improve its fit (through the "+ 2*k" term). The BIC is similar except that it also accounts for sample size and the penalty for increasing the number of parameters is typically more severe with BIC. Under the BIC, adding parameters has to lead to even larger improvements in the log likelihood in order for the larger model to be preferred.

[53] Note that this is a count of the number of IDS forms submitted, and not a count of how many document are cited on those IDS forms.

[54] The BIC and AIC are described in footnote 52. The BIC for Model 3 is equal to 1152.5, whereas the BIC for the model constrained to just the entity size, foreign origin and continuation history is equal to 1102.1. The model with the lower value for the BIC is preferred. The corresponding AICs are 1,034.6 and 1074. Again, the model with the lower AIC is preferred.

[55] Note that in the analyses in this section we did not generate a matched control group. Thus, we do not need to use a conditional logistic model.

[56] See MPEP 2141.

[57] Our results do not exclude instances in which the examiner cited prior art but did not use it as the basis for any rejection. However, this possibility is considered to be much less likely than failure to use prior art cited by the applicant.

[58] MPEP 609.01.

[59] http://www.uspto.gov/patents/init events/executive actions.jsp.

[60] As IT systems are updated, and legacy systems replaced, it will become easier to access and analyze these data.

[61] See, e.g., Robert Brendan Taylor, Burying, 19 Mich. Telecomm. & Tech. L. Rev. 99 (2012), available at http://www.mttl r.org/vol ni neteen/tayl or. pdf. 6http://www. https://www.federalregister.gov/regulations/0651-AB95/changes-to-information-disclosure-statement- requirements-and-other-related-matters. The proposed rule would have required applicant to provide an additional explanation of relevance if the reference were more than 25 pages long.

[62] https://www.federalregister.gov/regulations/0651-AB95/changes-to-information-disclosure-statement-requirements-and-other-related-matters. The proposed rule would have required applicant to provide an additional explanation of relevance if the reference were more than 25 pages long.

[63] The Cooperative Classification System (CPC) is a joint effort between the USPTO and the European Patent Office (EPO). The goal is to move toward a common classification scheme, thus enhancing the ability of examiners in both Europe and the United States to locate appropriate prior art patent documents. For more information, see http://www.uspto.gov/patents/resources/classification/index.jsp.

[64] http://w-pattr-05/stic/npl/index.cfm?type=About&var1=n/a.

[65] See http://www.uspto.gov/patents/init_events/preissuance_submissions.jsp.

[66] Crowdsourcing is one of a number of USPTO initiatives intended to strengthen patent quality. See http://www.uspto.gov/patents/init_events/crowdsourcing_roundtable_20141202.jsp.

[67] http://www.uspto.gov/patents/law/exam/interview_best_practices.pdf.

[68] http://www.uspto.gov/faq/pre_appealbrief_conf_pilot.jsp.

[69] http://www.uspto.gov/patents/init_events/faipp_landing.jsp/

[70] http://www.uspto.gov/news/pr/2014/14-08.jsp.

[71] KSR International Co. v. Teleflex Inc., 550 U.S. 398, 82 USPQ2d 1385 (2007); Bilski v. Kappos, 561 U.S. 593 (2010); Association for Molecular Pathology v. Myriad Genetics, Inc., 569 U.S. __, 133 S. Ct. 2107 (2013); Mayo Collaborative Serv. v. Prometheus Labs., Inc., 566 U.S. __, 132 S. Ct. 1289 (2012).

[72] See guidance documents available at http://www.uspto.gov/patents/law/exam/examguide.jsp.

[73] See, for example, http://www.uspto.gov/patents/law/index.jsp and pages linked therein.

[74] See, e.g., "Post Patent Issuance Challenges and the Quest for Patent Quality," Manny Schecter, September 11, 2014 http://www.ipwatchdog.com/2014/09/11/post-patent-issuance-challenges-and-the-quest-for-patent-quality/ and "Developing a patent portfolio that can withstand the 'death squad' for patents (Part 2)," Max Colice and Orion Armon, October 30, 2014 http://www.insidecounsel.com/2014/10/30/developing-a-patent-portfolio-that-can-withstand-t.

[75] Recall that the matching process attempts to match on value by using maintenance history and forward citations. However, that does not perfectly control for value.

[76] As mentioned in Section 3.3.2, one possible explanation is that many GS-13 examiners are in the Signatory Review Program and unlikely to allow claims unless they are clear and narrow in scope.

[77] Note that Cockburn et al. did find a positive correlation between the time allotted for examination and the likelihood of a validity ruling. Higher GS-level examiners generally have less time for each examination than lower GS-level examiners, so the result is not altogether inconsistent with the current study.

In: Patent Infringement Litigation
Editor: Lindsey Gonzales
ISBN: 978-1-63483-527-5

Chapter 2

INTELLECTUAL PROPERTY: ASSESSING FACTORS THAT AFFECT PATENT INFRINGEMENT LITIGATION COULD HELP IMPROVE PATENT QUALITY*

United States Government Accountability Office

WHY GAO DID THIS STUDY

Legal commentators, technology companies, Congress, and others have raised questions about patent infringement lawsuits by entities that own patents but do not make products. Such entities may include universities licensing patents developed by university research, companies focused on licensing patents they developed, or companies that buy patents from others for the purposes of asserting the patents for profit.

Section 34 of AIA mandated that GAO conduct a study on the consequences of patent litigation by NPEs. This report examines (1) the volume and characteristics of recent patent litigation activity; (2) views of stakeholders knowledgeable in patent litigation on key factors that have contributed to recent patent litigation; (3) what developments in the judicial system may affect patent litigation; and (4) what actions, if any, PTO has

* This is an edited, reformatted and augmented version of United States Government Accountability Office, Publication No. GAO-13-465, dated August 2013.

recently taken that may affect patent litigation in the future. GAO reviewed relevant laws, analyzed patent infringement litigation data from 2000 to 2011, and interviewed officials from PTO and knowledgeable stakeholders, including representatives of companies involved in patent litigation.

WHAT GAO RECOMMENDS

GAO recommends that PTO consider examining trends in patent infringement litigation and consider linking this information to internal patent examination data to improve patent quality and examination. PTO commented on a draft of this report and agreed with key findings and this recommendation.

WHAT GAO FOUND

From 2000 to 2010, the number of patent infringement lawsuits in the federal courts fluctuated slightly, and from 2010 to 2011, the number of such lawsuits increased by about a third. Some stakeholders GAO interviewed said that the increase in 2011 was most likely influenced by the anticipation of changes in the 2011 Leahy-Smith America Invents Act (AIA), which made several significant changes to the U.S. patent system, including limiting the number of defendants in a lawsuit, causing some plaintiffs that would have previously filed a single lawsuit with multiple defendants to break the lawsuit into multiple lawsuits. In addition, GAO's detailed analysis of a representative sample of 500 lawsuits from 2007 to 2011 shows that the number of overall defendants in patent infringement lawsuits increased by about 129 percent over this period. These data also show that companies that make products brought most of the lawsuits and that nonpracticing entities (NPE) brought about a fifth of all lawsuits. GAO's analysis of these data also found that lawsuits involving software-related patents accounted for about 89 percent of the increase in defendants over this period.

Stakeholders knowledgeable in patent litigation identified three key factors that likely contributed to many recent patent infringement lawsuits. First, several stakeholders GAO interviewed said that many such lawsuits are related to the prevalence of patents with unclear property rights; for example, several of these stakeholders noted that software-related patents often had overly broad or unclear claims or both. Second, some stakeholders said that the potential for large monetary awards from the courts, even for ideas that make only small contributions to a product, can be an incentive for patent owners to file infringement lawsuits. Third, several stakeholders said that the recognition by companies that patents are a more valuable asset than once assumed may have contributed to recent patent infringement lawsuits.

The judicial system is implementing new initiatives to improve the handling of patent cases in the federal courts, including (1) a patent pilot program, to encourage the enhancement of expertise in patent cases among district court judges, and (2) new rules in some federal court districts that are designed to reduce the time and expense of patent infringement litigation. Recent court decisions may also affect how monetary awards are calculated, among other things. Several stakeholders said that it is too early to tell what effect these initiatives will have on patent litigation.

The U.S. Patent and Trademark Office (PTO) has taken several recent actions that are likely to affect patent quality and litigation in the future, including agency initiatives and changes required by AIA. For example, in November 2011, PTO began working with the software industry to develop more uniform terminology for software-related patents. PTO officials said that they generally try to adapt to developments in patent law and industry to improve patent quality. However, the agency does not currently use information on patent litigation in initiating such actions; some PTO staff said that the types of patents involved in infringement litigation could be linked to PTO's internal data on the patent examination process, and a 2003 National Academies study showed that such analysis could be used to improve patent quality and examination by exposing patterns in the examination of patents that end up in court.

ABBREVIATIONS

AIA	Leahy-Smith America Invents Act
AIPLA	American Intellectual Property Law Association
AOUSC	Administrative Office of the U.S. Courts
CPC	Cooperative Patent Classification
EPO	European Patent Office
FTC	Federal Trade Commission
ITC	International Trade Commission
NPE	nonpracticing entities

PAE	patent assertion entities
PME	patent monetization entities
PTO	Patent and Trademark Office
R&D	research and development
SEC	Securities and Exchange Commission

* * *

August 22, 2013

Congressional Committees

History is filled with examples of successful inventors who did not develop products based on the technologies they patented. For example, Elias Howe patented a key component of the sewing machine—a mechanism for making a lockstitch—but it was Isaac Singer who most successfully brought these machines into the homes of thousands of Americans by obtaining crucial patents of his own and paying Howe and other inventors to license the technology described in their patents.[1] In the United States, the party that owns a patent—the patent owner—is not required to put the patent to use in order to profit from it; he can also license others to use it.[2] In some instances, patent owners may need to actively assert their patents in an adversarial context if another firm's product infringes their patents.[3] For example, Singer initially refused to obtain a license to Howe's patent, but when Howe sued Singer for infringing his patent, the two parties ultimately entered a licensing agreement.[4]

According to economists who have studied these issues, the U.S. patent system—authorized by the U.S. Constitution—aims to promote innovation by making it more profitable.[5] For example, a patent owner can generally exclude others from making, using, selling, or importing the patented technology for 20 years from the date on which the application for the patent was filed. By restricting competition, patents can allow their owners to earn greater profits on their patented technologies than they could earn if these technologies could be imitated freely. Due to the exclusive rights provided by patents, patents can help their owners recoup the costs of the research and development (R&D) of new technologies. On the other hand, any limiting effects on competition caused by the exclusive nature of patents may result in higher prices for products having patented technologies. The patent system, therefore, gives rise to complex tradeoffs involving innovation and competition. These trade-offs can be affected by decisions made by the United States Patent and Trademark

Office (PTO), which issues patents; the federal courts, which decide patent infringement lawsuits; and the International Trade Commission (ITC), which can order imports that infringe U.S. patents to be excluded from entering the country.

In addition to individual inventors who may choose not to develop products based on their patents, there are other types of “nonpracticing” patent owners, or nonpracticing entities (NPE). For example, some universities are NPEs, as they develop technologies in campus laboratories, and rather than producing and selling products that incorporate these technologies, they license their patents on these technologies to companies who use them in their products. In addition, some private firms are NPEs as they specialize in R&D, and rather than selling products, they license the patents for those products to fund further research. Some NPEs simply buy patents from others for the purpose of asserting them for profit; these NPEs are known as patent monetization entities (PME). 6 Other PMEs include companies that produced products at one time and still own patents on the technologies for those products. Experts agree that NPEs have a variety of business models, which makes it difficult to fit them neatly into any one of these categories. For example, even companies that produce products related to their patents—known as practicing patent owners, or operating companies—sometimes assert patents that they own but that are not related to the products they produce, which further complicates defining an NPE.

Some legal commentators, technology companies, the Federal Trade Commission (FTC),[7] and Congress, among others, have raised concerns that patent infringement litigation by NPEs is increasing and that this litigation, in some cases, has imposed high costs on firms that are actually developing and manufacturing products, especially in the software and technology sectors. Among the concerns of some technology companies and legal commentators is that because NPEs generally face lower litigation costs than those they are accusing of infringement, NPEs are likely to use the threat of imposing these costs as leverage in seeking infringement compensation.[8] These technology companies and legal commentators also have noted that NPEs often claim that their patent covers an entire technology when in fact it may cover just a small improvement in an existing technology, and that it can be difficult for judges and juries to determine the patent’s scope when complex technologies are involved.

The Leahy-Smith America Invents Act (AIA), signed into law September 16, 2011, made several significant changes to the U.S. patent system. Section 34 of the AIA 9 mandates that GAO conduct a study on the consequences of patent litigation by NPEs.[10] Our objectives in conducting this study were to determine: (1) what is known about the volume and characteristics of recent patent litigation activity; (2) the views of stakeholders knowledgeable in patent litigation on what is known about the key factors that have contributed to recent patent litigation; (3) what developments in the judicial system may affect patent litigation; and (4) what actions, if any, has PTO recently taken that may affect patent litigation in the future.

To address all four objectives, we reviewed relevant laws, including the AIA, and interviewed officials from PTO, FTC, ITC, and 44 stakeholders knowledgeable about patent litigation. These included representatives from companies and industry groups that were recently sued for patent infringement, PMEs, judges, various legal commentators (including law professors and patent litigators representing operating companies and PMEs), economists, representatives from research universities that license patents, patent brokers who help others buy and sell patents, venture capitalists, and individual inventors. 11 To describe what is known about the volume and characteristics of recent patent litigation activity for 2007 to 2011, we analyzed patent infringement litigation data from Lex Machina, a firm that collects and analyzes data on patent litigation. Lex Machina provided data for all patent infringement lawsuits filed in federal district court from 2000 to 2011. From these data, Lex Machina selected a random, generalizable sample of 500 lawsuits (100 per year from 2007 to 2011), which allows us to estimate percentages with a margin of error of no more than plus or minus 5 percentage points over all these years and no more than plus or minus 10 percentage points for any particular year.[12] Lex Machina used a variety of characteristics from court records, U.S. Securities and Exchange Commission (SEC) filings, and corporate websites to categorize litigants, including as an operating company or likely operating company, PME or likely PME, university, or an individual or trust.[13] A limitation of this categorization is that litigants were not contacted to verify their identity, so there is some uncertainty about the accuracy of the category in which Lex Machina placed them. We also obtained patent infringement data from RPX, another firm that collects data on patent infringement lawsuits, in an effort to verify Lex Machina's litigant categorizations.[14] Also to describe what is known about the volume and characteristics of recent patent litigation activity, we reported data collected by the American Intellectual Property Law Association (AIPLA) on the costs of patent litigation.[15] We also reviewed academic literature on patent litigation and the patent system in general and assessed the methodology of the studies we reported on for soundness. To assess the reliability of data from Lex

Machina, we met with Lex Machina staff, examined documentation, and tested and reviewed the data provided for completeness and accuracy. To assess the reliability of data from PTO, AIPLA, and RPX, we conducted interviews and reviewed relevant methodology documentation. We found these data to be sufficiently reliable for purposes of this report.

In addition to the steps we took to address all four objectives, in order to describe views of stakeholders on what is known about the key factors that contribute to recent patent litigation trends, we reviewed academic literature on the patent and judicial systems and the benefits and costs of patent assertion, including economic and legal studies. To describe developments in the judicial system that may affect patent litigation, we interviewed officials and judges from the U.S. District Courts for the District of Delaware and for the Eastern District of Texas. We selected these district courts because they had high levels of patent infringement lawsuits according to Lex Machina data. We also interviewed judges with the U.S. Court of Appeals for the Federal Circuit (Federal Circuit) in Washington, D.C., which hears appeals of patent cases decided in federal district courts, as well as officials from the Administrative Office of the U.S. Courts (AOUSC) and the Federal Judicial Center—organizations that provide broad administrative, legal, and technological services and support to the judicial branch. We also reviewed documents and data from the courts, as well as economic and legal studies. To describe what actions, if any, PTO has recently taken that may affect patent litigation in the future, we conducted interviews with officials from PTO and reviewed documents and data from the agency, as well as economic and legal studies. Appendix I provides more details on our scope and methodology.

We conducted this performance audit from November 2011 to August 2013 in accordance with generally accepted government auditing standards. Those standards require that we plan and perform the audit to obtain sufficient, appropriate evidence to provide a reasonable basis for our findings and conclusions based on our audit objectives. We believe that the evidence obtained provides a reasonable basis for our findings and conclusions based on our audit objectives.

BACKGROUND

When PTO receives a patent application, it assigns it to a team of patent examiners with relevant technology expertise. PTO does not begin examining patent applications upon receiving them and PTO's data shows that, as of June

2013, the average time between filing and an examiner's initial decision on the application was about 18 months.[16] On average, it takes 30 months for PTO to issue a patent once an application is submitted.

The focus of patent examination is determining whether the patent application satisfies the statutory requirements for a patent, including: (1) novelty, (2) nonobviousness, (3) utility, and (4) patentable subject matter.[17] Generally, other patents, publications, and publicly disclosed but unpatented inventions that pre-date the patent application's filing date are known as prior art. During patent examination, the examiner, among other things, compares an application's claims to the prior art to determine whether the claimed invention is novel and nonobvious.[18] The examiner then decides to reject or grant the claims in the application and deny the application or grant a patent.

U.S. patents include the specification and the claims:

- The specification is a written description of the invention that, among other things, sufficiently discloses the invention and the manner and process of making and using it. The specification must be written in full, clear, concise, and exact terms so as to enable any person skilled in the art to make and use the invention. As an example, an excerpt from the specification for a cardboard coffee cup and insulator invention describes "corrugated containers and container holders which can be made from existing cellulosic materials, such as paper."
- The claims define the scope of the invention for which protection is granted and must be definite. There are often a dozen or more claims per patent, and they can often be difficult for a layperson to understand, according to legal commentators. For example, one claim for the cardboard coffee cup insulator begins by referring to "a recyclable, insulating beverage container holder, comprising a corrugated tubular member comprising cellulosic material and at least a first opening therein for receiving and retaining a beverage container." A patent's claims can be written broadly or be more narrowly defined, according to legal commentators, and applicants can change the wording of claims—which can affect their scope—during examination based on examiner feedback. Patents are a property right and—like land—their claims define their boundaries. When a property right is not clearly defined, it can lead to boundary disputes, although to some extent uncertainty is inherent. Consequently, legal commentators define high-quality patents as

those whose claims clearly define and provide clear notice of their boundaries.

Once issued by PTO, a patent is presumed to be valid. However, the patentability of its claims can be challenged in administrative proceedings before PTO or its Patent Trial and Appeal Board and its validity can be challenged in federal court. For example, the AIA established three new administrative proceedings for entities to challenge the patentability of a patent's claims:

- Inter partes review.[19] This proceeding allows anyone who is not the patent owner to request review of an issued patent by presenting prior art to PTO—either patents or other publications—to challenge the claimed invention's patentability as obvious or not novel. This review proceeding became available on September 16, 2012, 1 year after the enactment of the AIA, but entities cannot request this review until the later of (1) 9 months after a patent is granted or (2) completion of post-grant review, if such a proceeding is held.
- Post-grant review. This proceeding allows anyone who is not the patent owner to request a review of an issued patent that challenges at least one of the claims' patentability in more circumstances than inter partes review—such as the invention not being useful—and not solely based on prior art. This proceeding is available for patents issued from patent applications with a filing date of March 16, 2013, or later and requests must be filed within 9 months of the patent's issuance.
- Transitional program for covered business method patents.[20] This proceeding allows anyone who is sued or charged with infringing a covered business method patent to request a review of the patent to challenge a claim's patentability in generally the same circumstances as post-grant review. This review proceeding became available on September 16, 2012, and requests must be filed within 9 months of the patent's issuance.

In addition, a patent's validity can be challenged in the 94 federal district courts throughout the country by, for example, presenting additional prior art that PTO may have been unaware of when it granted the patent. Challenges to a patent's validity are often brought by an accused infringer who has been sued for infringing the patent.[21] Patent owners can bring infringement lawsuits against anyone who uses, makes, sells, offers to sell, or imports the patented invention without authorization because a patent is a right to exclude others from practicing the invention. Exactly what a patent covers and whether another product infringes the patent's claims are rarely easy questions to resolve in litigation, according to legal commentators. As noted, appeals of district court decisions in infringement cases are heard in the U.S. Court of Appeals for the Federal Circuit in Washington D.C.

If a patent infringement lawsuit is not dismissed in the initial stages, it proceeds to discovery (a process that exists in all federal civil litigation) and claim construction. Discovery requires the accused infringer to produce documents or other information that shows, among other things, how the allegedly infringing product is made and operates to help the patent owner establish infringement. Similarly, the patent owner must produce documents or other information that the accused infringer can use to challenge the patent's validity, among other things. However, parties that do not offer products or services using the patents at issue often have far fewer documents to disclose—because they do not have any documents related to their products or services—than patent owners or accused infringers who do offer products or services.[22]

With this information the patent owner specifies which patent claims allegedly are infringed and the alleged infringer responds by explaining why the allegedly infringing product is not covered by the patent's claims. This identifies the patent claims the court needs to interpret. Known as claim construction, this is a fundamental issue in patent cases, and each party tries to persuade the court to interpret the patent claims in its favor. The court has broad discretion in how it goes about this process, which can involve a hearing with testimony from witnesses, according to legal commentators.[23] In addition, if the patent's validity is being challenged, the alleged infringer specifies why the patent allegedly is not valid, including any prior art.

Once the judge interprets the claims, the claims are then applied to the allegedly infringing product, to determine infringement, and to the prior art, to determine the patent's validity if it is challenged. If the patent is found to be both valid and infringed, the court can award the patent owner monetary damages, issue an injunction to prohibit further infringement, or both. The court is required to award damages adequate to compensate for the infringement that are at least what a reasonable royalty would be for the use made of the invention by the infringer.[24]

In addition to being enforced in the federal courts, patents can also be enforced at ITC, which handles investigations into allegations of certain unfair practices in import trade. Specifically, certain patent owners can file a complaint with ITC if imported goods infringe their patent or are made by a process covered by the patent's claims.[25] If ITC determines after an investigation that an imported good infringes a patent, the agency can issue an exclusion order barring the products at issue from entry into the United States, which the President can disapprove for policy reasons. ITC decisions can be appealed to the U.S. Court of Appeals for the Federal Circuit. Legal commentators have reported that a 2006 Supreme Court decision led to increased complaints alleging imported goods infringed U.S. patents being filed with ITC, and recent ITC data show that the number of investigations instituted by ITC increased from 32 in 2006 to 37 in 2012.[26]

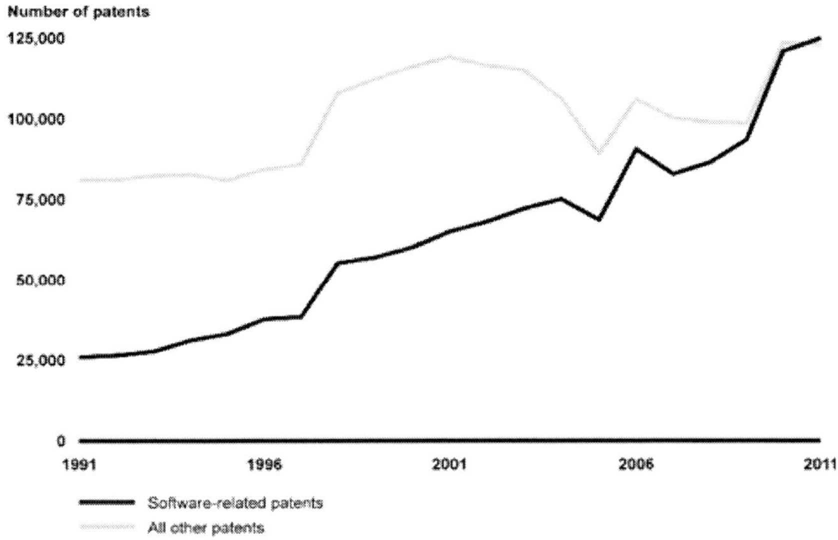

Source: GAO analysis of United States Patent and Trademark Office data.

Note: Software-related patents include a number of patent classes that are most likely to include patents with software-related claims, and this includes business method patents.

Figure 1. Number of Software-Related Patents Granted per Year by PTO, 1991 to 2011.

According to PTO data, applications for all types of patents have increased in recent years, and patents granted for software-related technologies have seen dramatic increases over the past 2 decades (see fig. 1).[27] Software-related patents occur in a variety of technologies containing at least some element of software, and cover things like sending messages or conducting business over the Internet (e.g. e-commerce). Patents related to software can, but do not generally, detail computer software programming code in the specification, but often provide a more general description of the invention, which can be programmed in a variety of ways.

According to legal commentators, the number of software-related patents grew as computers were integrated into a greater expanse of everyday products. By 2011 patents related to software made up more than half of all issued patents. Software was not always patentable, and Supreme Court decisions in the 1970s found mathematical formulas used by computers were not patentable subject matter.[28] However, a 1981 Supreme Court decision overturned PTO's denial of a patent application for a mathematical formula and a programmed digital computer because, as a process, it was patentable subject matter.[29] Subsequently, in 1998, the Federal Circuit ruled that a mathematical formula in the form of a computer program is patentable if it is applied in a useful way.[30] According to PTO officials, the agency interpreted these cases as limiting their ability to reject patent applications for computer processes. Legal commentators also said that after these decisions, particularly the 1998 Federal Circuit decision, software-related patenting grew as many technology companies made the conscious effort to generate more patents for offensive or defensive purposes—that is, to use them to sue or countersue competitors in infringement lawsuits, rather than use them to recoup R&D costs. As recently as 2010, the Supreme Court has noted that the patent statute acknowledges that business methods are patentable subject matter.[31]

NUMBER OF PATENT INFRINGEMENT LAWSUITS INCREASED SIGNIFICANTLY IN 2011 AND THE NUMBER OF DEFENDANTS INCREASED BETWEEN 2007 AND 2011

From 2000 to 2010, the number of patent infringement lawsuits fluctuated slightly, and from 2010 to 2011, the number increased about 31 percent. Our more detailed analysis of a generalizable sample of 500 lawsuits estimates that the overall number of defendants in these cases increased from 2007 to 2011 by about 129 percent over the 5-year period. This analysis also shows that operating companies brought most of these lawsuits and that lawsuits involving software-related patents accounted for about 89 percent of the increase in defendants during this period. Some stakeholders we interviewed said that they experienced a substantial amount of patent assertion without firms ever filing lawsuits against them.

Number of Patent Infringement Lawsuits Fluctuated Slightly before Increasing in 2011, but Number of Defendants Increased from 2007 to 2011

From 2000 to 2011, about 29,000 patent infringement lawsuits were filed in U.S. district courts. The number of these lawsuits fluctuated slightly until 2011, when there was a 31 percent increase (see fig. 2).

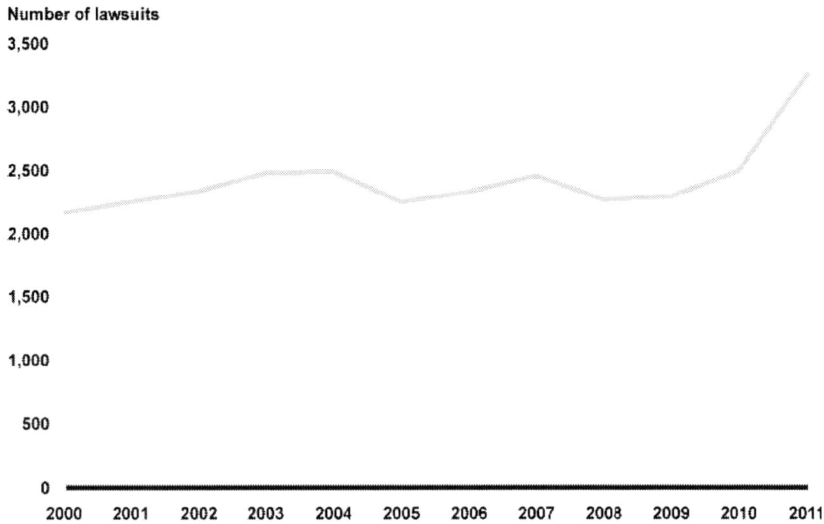

Source: GAO analysis of Lex Machina data.

Figure 2. Patent Infringement Lawsuits, 2000 to 2011.

Specifically, about 900 more lawsuits were filed in 2011 than the average number of lawsuits filed in each of the previous years. Some stakeholders we interviewed generally attributed the increase in 2011 to patent owners' anticipation of the passage of AIA, which restricts the number of accused infringers who can be joined in a single lawsuit.[32] Prior to the enactment of AIA, plaintiffs could sue numerous defendants in a single lawsuit. AIA restricts this practice by prohibiting joining unrelated defendants into a single lawsuit based solely on allegations that they have infringed the same patent. According to the legislative history of AIA, this provision was designed to address problems created by plaintiffs joining defendants, sometimes numbering in the dozens.[33] As a result, some stakeholders we interviewed generally agreed that the increase in 2011 was due to the fact that plaintiffs had to file more lawsuits at the end of 2011 after AIA's enactment in order to sue the same number of defendants or anticipated this change and rushed to file lawsuits against multiple defendants before it was enacted. In addition, our analysis of a generalizable sample of data from 2007 through 2011 estimates that the number of overall defendants in patent infringement suits increased by about 129 percent over the 5- year period (see fig. 3).

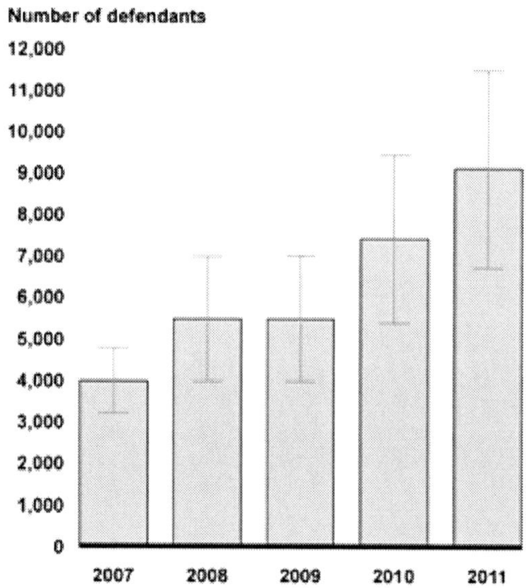

Source: GAO analysis of Lex Machina data.

Note: Defendant estimates are representative of all patent infringement lawsuits and error bars display 95 percent confidence intervals.

Figure 3. Estimated Number of Defendants in Patent Infringement Lawsuits, 2007 to 2011.

Representatives of several operating companies that we interviewed said they are being sued more often since the mid-2000s. For example, one former official at a large technology company told us that, in 2002, the company was a defendant in five patent infringement lawsuits, but in 2011, it was a defendant in more than 50. However, a few legal commentators we interviewed said that such increases are common during periods of rapid technological change—new industries lead to more patents and the number of patent infringement lawsuits also increases because there are more patents to be enforced. Similarly, one researcher working on these issues told us that, historically, major technological developments—such as the development of automobiles, airplanes, and radio—have also led to temporary, dramatic increases in patent infringement lawsuits.

Characteristics of Plaintiffs in Recent Patent Litigation

Operating companies brought most of the patent infringement lawsuits from 2007 to 2011. According to our analysis of data for this period, operating companies and related entities brought an estimated 68 percent of all lawsuits.[34] PMEs and likely PMEs brought 19 percent of the lawsuits.[35] PMEs and likely PMEs brought 17 percent of all lawsuits in 2007 and 24 percent in 2011, although this increase was not statistically significant. In contrast, operating companies and related entities filed 76 percent of the lawsuits in 2007 and 59 percent in 2011, a statistically significant decrease.[36] Individual inventors brought about 8 percent of the lawsuits, and research firms and universities brought less than 3 percent over the 5 year span. In about 3 percent of the lawsuits there was insufficient evidence to determine the type of plaintiff (see fig. 4).

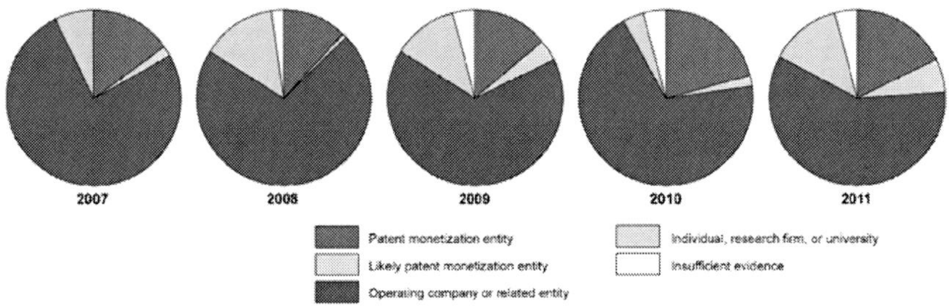

Source: GAO analysis of Lex Machina data.

Note: Lawsuit estimates are subject to a margin of error of up to plus or minus 10 percentage points.

Figure 4. Estimated Patent Infringement Lawsuits by Type of Plaintiff, 2007 to 2011.

Our analysis of the data from 2007 through 2011 shows that PMEs tended to sue more defendants per suit than operating companies. For this period, there were about 1.9 defendants on average for suits filed by operating companies, and about 4.1 defendants on average for suits filed by PMEs. In addition, a disproportionate share of PMEs sued a relatively large number of defendants. For example, about 12 percent of PMEs sued 10 defendants or more in a single lawsuit, compared to about 3 percent of operating companies, a statistically significant difference. Thus, even with bringing about a fifth of all patent infringement lawsuits from 2007 to 2011, PMEs sued close to one-third of the overall defendants, accounting for about half of the overall increase in defendants. Additionally, the estimated total number of defendants sued by PMEs more than tripled from 834 in 2007 to 3,401 in 2011, while the increase in the total number of defendants sued by operating companies was not statistically significant.[37]

To further explore the characteristics of PMEs and other NPEs, we interviewed stakeholders knowledgeable about patent infringement lawsuits. According to some stakeholders we interviewed, NPEs have many different characteristics and there are a spectrum of NPE business models and behaviors, including operating companies that partner with NPEs to file infringement suits.[38] These different types of NPEs include the following:

- PMEs and likely PMEs. PMEs we spoke with did not develop technology or sell products but, instead, derived most of their revenue from asserting patents against operating companies. Some of these PMEs told us that they acquired patents from a variety of sellers, such as universities, individual inventors, failed companies, or operating companies. A few of these PMEs told us they were able to get patents on their own even with minimal R&D investments, especially for software-related processes. Some PMEs we spoke with said that they formerly produced patented products and now simply assert those patents, and others said that they sued on behalf of individual inventors who did not have the resources to enforce patents on their own.
- Entities Related to Operating Companies. Our analysis of patent infringement lawsuit data shows that some entities were subsidiaries of or had other corporate relationships with operating companies, although they did not produce products themselves. In addition, some stakeholders we interviewed said that operating companies sometimes partner with PMEs to monetize patents. In some cases, these partnerships may allow an operating company to sue its competitors with less risk of countersuits. A few operating companies we spoke

with acknowledged that they are aware of such partnerships, although none said they engaged in this practice. The two patent brokers we interviewed told us that they have structured agreements where they transfer patents from an operating company to a PME, and one legal commentator who also structured such deals told us that operating companies often retain an interest in the gains from any lawsuits the PME files. This commentator also said that there are many PME lawsuits in which the identity of interested operating companies is intentionally hidden; our review of court records indicates that corporate relationships may not be easily deciphered. In some cases, business relationships were easily identifiable from court records and in other cases the links were more difficult to identify. [39] Further, some operating companies, like PMEs, also assert patents not related to products they produce, according to some stakeholders.[40]

- Research firms. Representatives from a few of the companies we interviewed said that they invested heavily in R&D and made efforts to share their technology with other companies and to help them develop new products. Specifically, these representatives told us that their companies did not focus on producing products but, rather, mainly developed new technologies and then licensed them to operating companies to pay for continued R&D.[41] These representatives also said that they provided technical support to the firms they license patents to helping them to make the best use of their patented technologies, which distinguishes them from PMEs.
- Universities. Many universities license their patented technologies to companies who use them in their products, according to a representative from each of two large research universities we spoke with, although they said that the licensing revenue is generally small in relation to other sources of university revenue. They also noted that licensing at many universities is mostly driven by life sciences research and that, sometimes before research begins, universities develop partnerships with private sector firms, such as pharmaceutical companies, who have the first right to bring patented ideas into the marketplace.

Types of Patents Involved in Recent Patent Infringement Litigation

Our analysis of patent infringement lawsuit data from 2007 to 2011 shows that on average about 46 percent of the lawsuits involved software-related patents. Between 2007 to 2011, 64 percent of defendants were sued over software-related patents, and these patents were at issue in the lawsuits that accounted for about 89 percent of the increase in defendants over this period (see fig. 5). About 49 percent of the patents in our sample were asserted within 5 years of being granted, and there was no statistically significant difference between software-related patents and other patents in this regard.

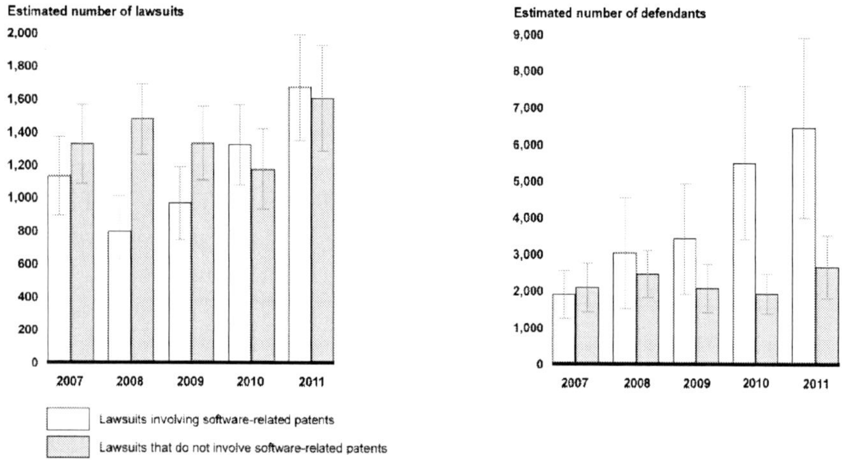

Source: GAO analysis of Lex Machina data.

Notes: Lawsuit and defendant estimates are representative of all patent infringement lawsuits and error bars display 95 percent confidence intervals. Software-related patents include a number of patent classes that are most likely to include patents with software-related claims, and this includes business method patents.

Figure 5. Estimated Number of Patent Infringement Lawsuits and Defendants Associated with Software-Related Patents, 2007 to 2011.

Our analysis of patent infringement lawsuit data from 2007 to 2011 shows that operating companies and PMEs both asserted software-related patents, although PME lawsuits involved these patents to a much greater extent. Specifically, about 84 percent of PME lawsuits from 2007 to 2011 involved software-related patents, while about 35 percent of operating company lawsuits did. However, operating companies brought a greater number of lawsuits involving software-related patents, given that they filed more lawsuits overall.[42] By defendant, software-related patents were used to sue 93 percent of the defendants in PME suits and 46 percent of the defendants in operating company suits (see fig. 6).

Technology-related operating companies were not the only companies sued for infringing software-related patents; other sectors were also sued for infringing such patents, including retail companies and local governments. We estimate that 39 percent of suits involving software-related patents were against firms in nontechnology sectors, according to our analysis of 2007 to 2011 data. One representative from a retail company noted that historically, all of the patent infringement lawsuits brought against the company used to be related to products they sold. However, as of mid-2012, the representative said that half of the lawsuits against the company were related to e-commerce software that the company uses for its shopping website—such as software that allows customers to locate their stores on the website—and were brought by PMEs. Representatives of retail and pharmaceutical companies told us they also defend lawsuits brought by PMEs related to features on their websites—typically software that outside vendors provide to them, rather than something they developed.

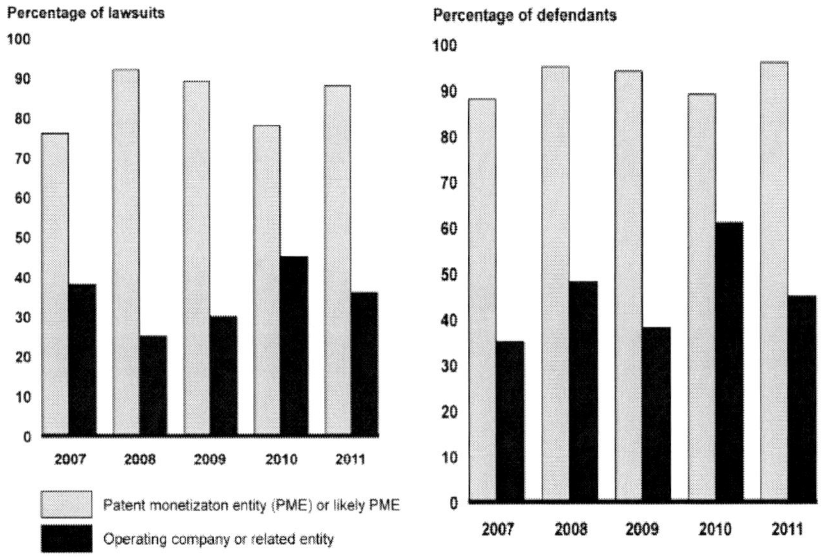

Note: Percentage of lawsuit and defendant estimates are subject to a margin of error of up to plus or minus 10 percentage points. Software-related patents include a number of patent classes that are most likely to include patents with software-related claims, and this includes business method patents. PMEs focus solely on asserting typically purchased patents.

Figure 6. Estimated Percentage of PME and Operating Company Lawsuits and Defendants Associated with Software-Related Patents, 2007 to 2011.

Additionally, city public transit agencies have been sued for allegedly infringing patents by using software for real-time public transit arrival notifications, according to a few stakeholders we interviewed.

Common Venues of Recent Patent Litigation

For 2007 to 2011, an estimated 32 percent of patent infringement lawsuits were filed in 3 of the 94 federal district courts: the Eastern District of Texas, the District of Delaware, and the Central District of California. These districts also had the most lawsuits filed for the period of 2000 to 2011 (see fig. 7).

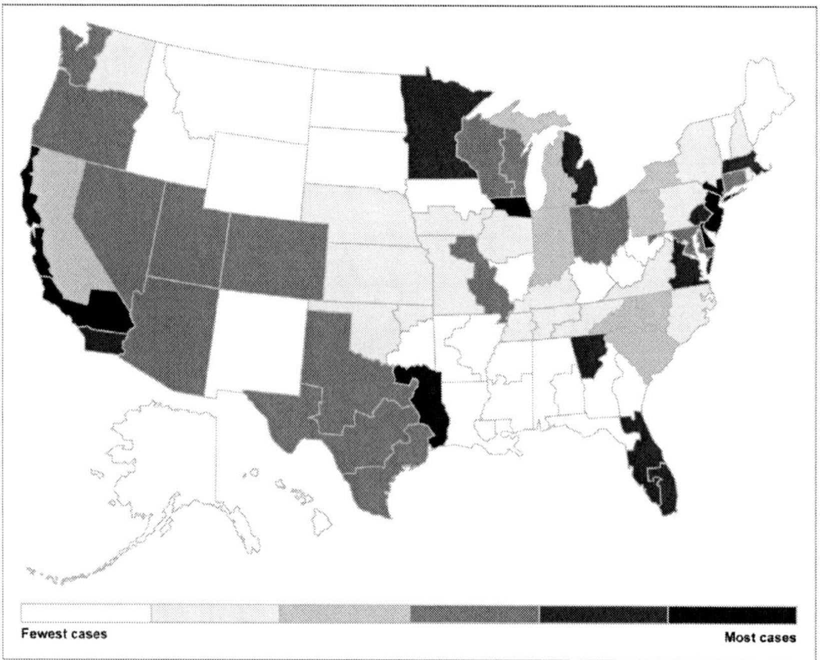

Source: GAO analysis of Lex Machina data; Map Resources (map).

Figure 7. Distribution of Total Patent Infringement Lawsuits across U.S. District Courts from 2000 to 2011.

In addition, our review of 2007 to 2011 litigation data shows that PMEs filed more lawsuits in the Eastern District of Texas than other types of plaintiffs. Specifically, from 2007 to 2011, 39 percent of PME and likely PME lawsuits were filed in the Eastern District of Texas, compared to about 8 percent of lawsuits filed by all other plaintiffs. Some stakeholders we interviewed said that in their view this occurs because juries in this district favor patent owners over alleged infringers.[43] In addition, one study we reviewed that looked at all decisions for all patent lawsuits from 1995 to 2011, showed that the Eastern District of Texas, the Eastern District of Virginia, and the District of Delaware were among the top districts for quicker trials, higher success rates, and higher damage awards for patent owners.[44]

Outcomes and Costs of Recent Patent Litigation

An estimated 21 percent of patent infringement cases from 2007 to 2011 were still ongoing. Of the remaining cases, our analysis shows that about 86 percent either ended or likely ended in a settlement. This occurred because both parties agreed to a judgment that the judge sanctioned, both parties agreed to end the lawsuit, or the plaintiff, who had brought the lawsuit, asked for it to be dismissed.[45]

Lawsuits brought by both operating companies and PMEs settled or likely settled at similar rates.[46] Less than 3 percent of the cases that were not ongoing ended in a trial and judgment, or were on appeal, which was consistent with what some representatives of operating companies told us—very few of their lawsuits go to trial because they settle quickly to avoid high litigation costs.

We were not able to determine litigation cost information from our sample data, and we found very little information on the costs of patent infringement lawsuits in court records. Further, as one stakeholder we interviewed noted, all litigation is expensive, not just patent infringement litigation. According to a 2011 nongeneralizable survey of patent lawyers by AIPLA, the cost of defending one patent infringement lawsuit, which excludes any damages awarded, was from $650,000 to $5 million in 2011, depending on how much was at risk.[47]

As for damages awarded, a 2012 study that looked at all district court patent decisions that proceeded through trial from 1995 to 2011 found that the median damage award was over $5 million dollars and that damage awards in NPE cases were higher than in other types of suits.[48] The authors of a 2012 paper who collected data from a nonrandom, nongeneralizable set of 82 operating companies noted that total litigation costs for NPE suits, including damages awarded and legal fees, were around $300,000 for small and medium companies and $600,000 for large companies.[49] The author of another 2012 paper sought to examine the impacts of NPE litigation on small companies and collected data from a nonrandom, nongeneralizable set of 223 small technology companies.[50] Of the 79 companies that indicated that they had received a patent demand, 31 reported that the demand affected the company in various ways, including reduced hiring or a reduced value of their company—which the author collectively described as "significant operational impacts."

Stakeholders Reported That Patent Assertion Occurs without Firms Ever Filing Lawsuits

In addition to lawsuits, patent assertion occurs without firms ever filing lawsuits, but the extent of this practice is unclear because we were not able to find reliable data on patent assertion outside of the court system. According to representatives of some operating companies we spoke with, they often get letters from patent owners offering licenses for the use of their patents. They said that these letters, which they refer to as "demand letters," sometimes threaten lawsuits if the parties do not reach a licensing agreement. These company representatives told us that for every patent infringement lawsuit filed against them, they might receive many times more letters notifying them of potential infringement and offering licenses.[51]

Representatives from a few operating companies we spoke with said that these letters can sometimes help to resolve issues without litigation, but that at times the letters can be so vague that they do not reference the patents at issue or what products the operating company sells that may be infringing these patents.

However, a few PME representatives told us that operating companies generally ignore their letters, thus leading the PMEs to sue the companies first to get their attention.

A few of these PMEs also told us that they are more likely to sue without sending a demand letter after a 2007 Supreme Court decision expanded accused infringers' ability to file preemptive declaratory judgment lawsuits seeking determinations that the patent is invalid, unenforceable, or not being infringed. 52 The threat of a declaratory judgment lawsuit can derail patent owners' attempts to reach a licensing agreement, according to a few PMEs we spoke with.

Because licenses or payments resulting from out-of-court patent assertions are almost always confidential, it is difficult to know the cost of these settlements.

The authors of the 2012 study noted above of a nonrandom, nongeneralizable set of 82 operating companies attempted to identify this cost. The 46 companies that provided data on costs reported that they spent an average of about $30 million on NPE suits, including both legal fees and settlements, which were settled without litigation from 2005 to 2011.

Stakeholders Identified Three Key Factors Contributing to Many Recent Patent Infringement Lawsuits

Stakeholders we spoke with identified three key factors that likely contributed to many recent patent infringement lawsuits: (1) unclear and overly broad patents, (2) the potential for disproportionately large damage awards, and (3) the increasing recognition that patents are a valuable asset.

Stakeholders Said That Some Patents Have Unclear Property Rights and Make Overly Broad Claims

Several of the stakeholders we spoke with, including representatives from PMEs, operating companies, and legal commentators, said that many recent patent infringement lawsuits are related to the prevalence of low-quality patents; that is, patents with unclear property rights, overly broad claims, or both. Although there is some inherent uncertainty associated with all patent claims, several of the stakeholders with this opinion noted that claims in software-related patents are often overly broad, unclear or both. Unclear and overly broad patents do not provide notice about their boundaries and the uncertainty of a patent's scope then usually needs to be resolved in court, according to some stakeholders we spoke with.

Stakeholders we interviewed identified several reasons why patents may be overly broad, unclear, or vague:

- Some stakeholders representing different interests, including operating companies, PMEs, and legal commentators, said the use of unclear terminology in patents can lead to a lack of understanding of patent claims and, therefore, what constitutes infringement, which needs to be resolved in court. For example, two of these stakeholders said the computer software industry does not have clear terminology or common vocabularies for describing concepts, innovations, and ideas. Language describing emerging technologies, such as software, may be inherently imprecise because these technologies are constantly evolving. In contrast, pharmaceutical drug patents are relatively clear because they use standardized scientific terminology, according to a few stakeholders.

- Some stakeholders, including operating companies and legal commentators, emphasized that claims in software patents sometimes define the scope of the invention by encompassing an entire function–-like sending an e-mail—rather than the specific means of performing that function.[53] According to a few stakeholders we spoke with, functional claims and other overly broad claims may allow patent owners who sue for infringement to argue in court that their patent covers (1) an entire technology when it may only cover a small improvement, or (2) future technologies that their patent did not originally intend to cover. For example, representatives from one PME we spoke with said they had successfully sued companies for infringement even though the companies were implementing their idea in a completely different manner than described in their patent—noting that they had patented their invention before the technology to best implement it was actually available. In addition, a few stakeholders noted that in certain circumstances patent owners can file a reissue application with PTO, which can sometimes broaden the scope of patents after they are issued.[54] In addition, a few stakeholders noted that patent owners can also file continuation applications during patent examination, which, depending on the type filed, allow new subject matter to be added to the patent application.[55]
- According to some stakeholders—including PMEs and representatives of operating companies that were sued for patent infringement—broad patents on concepts they would not expect to be patented make it easy to infringe a patent without intending to do so, although this is not a defense since patent infringement is a strict liability offense.[56] For example, the software start-up and venture capital firms we spoke with said that many software-related patents they encountered were for obvious concepts and did not represent any real contribution to new technology. Other stakeholders, including operating companies, judges, and legal commentators, also said some patents, particularly software-related patents, should never have been issued because they were obvious, not novel, or lacked definiteness.

Unclear boundaries make it hard to determine whether a patent is actually related to a particular technology. Several diverse stakeholders, including PMEs, operating companies, legal commentators, and judges we interviewed said that many overly broad or vague patent claims do not sufficiently identify the scope of the patent's coverage. This lack of notice makes it difficult for entities to identify relevant existing patents and prior art before developing new products, according to some legal commentators and operating companies. This difficulty reduces the effectiveness of searching for existing patents to ensure the product being developed does not infringe on valid patents and, according to several stakeholders, including PMEs, legal commentators, and venture capital and start-up firms, entities do not always conduct patent searches. For example, representatives from a software start-up company we spoke with told us that searching for relevant patents before developing new products is unrealistic and diverts already scarce resources, particularly because their product development process can be as short as 2 months. In contrast, pharmaceutical company representatives we spoke with said that development of new drugs is so expensive that patent searches are a necessity and that they can conduct thorough searches because their patents are described clearly and are easy to find in industry databases.

In addition, a few stakeholders we spoke with told us that the sheer volume of patents makes searching for relevant patents before developing new products particularly difficult, especially for products that combine many patented technologies. For example, these stakeholders we spoke with estimated that a typical smartphone uses from 50,000 to 250,000 patented technologies because such devices incorporate technologies from digital cameras, global positioning systems, and wireless communication. In contrast, pharmaceutical company representatives we spoke with said they are able to conduct thorough patent searches, in part, because there are fewer patents per drug. Determining the licenses a new product needs can be costly when many patents are involved; according to data from the AIPLA survey, the median cost for a legal validity and infringement opinion is $15,000 per patent.[57] Additionally, a technology start-up company told us that they may avoid patent searches because damage awards can be tripled for willful infringement, and by not searching for existing patents, they can claim ignorance.

Even if an entity conducts a patent search, identifies a relevant patent, and wants to avoid infringing it by obtaining a license to use it, according to several diverse stakeholders we spoke with, including PMEs, legal commentators, and operating companies, identifying the owner could be difficult because patent owners are not required to notify PTO of changes in assignment or ownership. According to some stakeholders we spoke with, finding patent owners is further complicated by the fact that some of these patent owners create subsidiary companies solely to hold their patent rights. In fact, several stakeholders we interviewed, including legal commentators, operating companies, and a PME, said that some entities intentionally hide the existence of their patents until a sector or company are using the patented invention without authorization and can be sued for infringement. Further, some economic literature we reviewed suggests that the numerous technologies in many products are sometimes patented by many different patent owners and can have overlapping rights, making it difficult and costly to determine which patents the operating company needs to license. We discuss PTO's current efforts to address these notice problems later in the report.

Most of the representatives from operating companies we interviewed said that PMEs specifically have played a role in the rise in patent infringement litigation. Some of the representatives from operating companies also said that PMEs are often more willing to bring lawsuits based on a broad interpretation of their patents' claims because they cannot be countersued for patent infringement since they do not produce a product. Some economic literature we reviewed suggests that accused infringers have an incentive to settle quickly to avoid the uncertainty of claim construction and high litigation costs, particularly if they face very high costs of changing their products to avoid infringement.[58] Although a few stakeholders said that operating companies have also brought lawsuits alleging infringement of poor quality patents that they do not practice, and several stakeholders, including PMEs, operating companies, and legal commentators, said they believe the prevalence of low quality patents was driving recent increases in litigation more than PME suits.[59]

Stakeholders Believe the Potential for Disproportionately Large Damage Awards from the Courts Can Incentivize Litigation

Disproportionately large damages awarded by the courts can motivate patent owners to file lawsuits, according to several stakeholders, we spoke with, including operating companies, PMEs, and legal commentators. According to some stakeholders, the potential for large damage awards may encourage some patent owners to file lawsuits in the hope that the accused infringer settles to avoid going to court. A few representatives of operating companies told us that companies prefer to settle lawsuits before trial for smaller amounts of money rather than risk having to pay large damage awards and legal fees, even if they know the case against them is weak. They said that some patent owners file infringement lawsuits knowing that defendants will settle the case before the court determines whether infringement occurred or the patents are valid. In addition, some operating company representatives said that patent owners often sue operating companies that have purchased licenses to use software in order to get settlements from numerous infringers rather than suing the vendor. One PME we spoke with said that although it tries to sue technology vendors whenever possible, it sues end users most of the time because these are usually profitable companies, and there is greater potential for larger settlements.

Until 2011, damage awards could be calculated using the 25 percent rule—whereby the alleged infringer would pay a royalty rate equivalent to 25 percent of the expected profits for the product that incorporates the patent at issue. As a result, according to some stakeholders, damage awards were outsized and did not reflect the value of the patent or the patent's contribution to the product at issue, making it possible for the owner of only a few patents to capture most of the profit from a product, even though thousands of other patents also contributed to the development of the product. In a 2011 report, FTC noted that some legal rules for calculating damages were not grounded in economic analysis and therefore may under- or overcompensate patent owners for infringement in comparison to the market.[60] FTC reported that overcompensation can overincentivize patenting and patent litigation. Moreover, the FTC report included concerns from some industry participants that the value added by one patented technology may be very small relative to the price of the entire product for complex products.

In its 2011 report, FTC made a number of recommendations to courts, including that damage awards—since they are based on reasonable royalties—should not be higher than the incremental value of the patented invention over the next best alternative and should not be punitive. Some stakeholders noted that changes to damage awards occurred after a 2011 Federal Circuit decision eliminated the use of the 25 percent rule because it was fundamentally flawed and established a rule of apportionment whereby damages are apportioned according to the patent at issue's contribution to the product. 61 These stakeholders noted this 2011 decision and other decisions have had the effect of lowering damage awards to reflect a patent's value or eliminated the potential for enhanced damages.[62] Despite some of this progress, a few stakeholders noted that there is still a need for improvement in the way damages are calculated.

Stakeholders Noted the Recognition of Patents As a Valuable Asset May Have Contributed to Recent Patent Issuance and Litigation

Several stakeholders, including PMEs and legal commentators, we interviewed said that the recognition that patents are a more valuable asset than once assumed may have contributed to recent patent issuance trends and patent infringement lawsuits. Within the last 10 years, technology companies in particular have increasingly realized that patents are valuable and can be important to their corporate strategy, according to some of these stakeholders. This trend may have started, according to literature we reviewed, when Texas Instruments Inc. was looking for additional sources of revenue in the 1980s and started to more aggressively assert its patents to increase revenue.[63] Prior to this, entities used patents to protect inventions rather than to generate revenue, according to some stakeholders, including legal commentators and a PME, we spoke with.

Despite the new value placed on patents, some industries may have slightly different approaches to patenting. For example, a few representatives of venture capital and software start-up firms told us that they do not always apply for patents until their companies are well established because patent attorneys are expensive, and the process is time consuming. They told us that the cost of R&D is low relative to the cost of applying for a patent, so there is minimal incentive in the software industry to patent in order to recoup R&D costs. For example, representatives from one small software company we spoke with said that they could develop a product in a little as 2 months with only a few programmers. They also noted that it is relatively easy to get patents on their software even with their minimal R&D efforts—it is just a matter of paying the patent attorney's fees. On the other hand, several representatives from the pharmaceutical industry told us that patents are actually critical to their ability to recoup the costs of developing a new drug, which can cost as much as $1 billion and take from 10 to 15 years.

Further, several stakeholders, including operating companies, PMEs, and legal commentators, we interviewed said that investors have played a role in the increasing number of patent infringement lawsuits. Specifically, according to some of these stakeholders, some PMEs have financial backers who fund the purchase and enforcement of patents and expect to see a return on their investment, sometimes turning to contingency fee law firms to carry out the lawsuits, where the law firm pays all of the litigation costs but shares in any award or settlement.[64]

Furthermore, economic literature suggests that inventors who do not have the resources or skills to enforce patents on their own benefit from partnering with PMEs that specialize in patent monetization, and this was confirmed by some of the stakeholders we spoke with. For example, one inventor we spoke with said that he was able to sell his patents to a PME that specialized in patent litigation when his start-up company failed, which allowed him to fund a new company. Representatives from a university we spoke with also said universities look to outside entities, such as PMEs, to finance patent infringement litigation because universities cannot cover the up-front costs of filing a lawsuit. Representatives from one PME we spoke with said that they helped a small inventor get the attention of large technology companies that were infringing his patent and ignoring his licensing requests. In addition, one patent broker we spoke with said that well-known operating companies often do not want to file patent infringement suits because of potential public backlash, so they sell or transfer their patents but retain an interest in any settlement or award.

NEW INITIATIVES IN THE COURTS MAY AFFECT PATENT LITIGATION

The federal court system is implementing some new initiatives to improve the handling of patent cases. A pilot program was established for certain U.S. district courts to encourage the enhancement of expertise in patent cases among district court judges and a model order was issued in response to high discovery costs for e-mails and other electronic documents in patent cases—known as e-discovery. In addition, some recent decisions in the Federal Circuit and the Supreme Court may also affect future patent infringement litigation.

Efforts to Improve Courts' Handling of Patent Cases

In January 2011, Congress established a pilot program in certain U.S. district courts to encourage the enhancement of expertise in patent cases among district court judges.[65] This pilot program would create a cadre of judges who have advanced knowledge of patents due to more intensified experience in handling the cases, according to a statement made during the congressional debate for the law. Specifically, the law required AOUSC to designate at least 6 district courts that met certain eligibility requirements to participate in a 10-year pilot program.[66] As of December 2012, AOUSC had designated all 14 federal district courts that applied as participants. Currently, 7 of the 14 district courts participating in the pilot program have adopted case management rules to govern patent cases, such as how cases are allocated to judges in the program, although some districts' rules predated the program.[67]

Some legal commentators and other stakeholders we interviewed said they were hopeful that the pilot program would lead to some meaningful improvements in patent case management in the courts. However, a few judges we interviewed said that more resources are needed to improve the handling of patent cases and that the needed resources were not appropriated to implement the pilot program.[68] For example, they told us that hiring additional law clerks would help with the increased workload and processing of patent cases resulting from participation in the pilot program. Because participation in the pilot program does not come with additional resources, some districts decided not to participate, according to the judges we spoke with. For example, a district court judge told us that his district court did not volunteer for the pilot program because funds were not appropriated to implement it. In general, several stakeholders we spoke with agreed that it is too early to tell what impact the patent pilot project will ultimately have on patent litigation.

Efforts to Reduce Discovery Costs

In response to high discovery costs, in September 2011, the Advisory Council for the U.S. Court of Appeals for the Federal Circuit issued a model order regarding discovery for e-mail and other electronic documents (known as e-discovery) targeted to patent cases.[69] According to some legal commentators and judges we interviewed, the technical complexity of patent cases leads to expansive discovery requests that are time consuming and expensive and, as is typically the case with most documents produced in discovery, do not necessarily produce documents used at trial. For example, a study conducted by AIPLA found that discovery costs can range from $350,000 to $3 million, depending on the size of the potential settlement.[70]

One judge we spoke to said that only a few of the documents in discovery are actually used at trial—often less than one document in 10,000—and representatives from one operating company told us that about 2,000 of the 10 million documents they were asked to produce were actually used in a recent trial.

After the issuance of the e-discovery model order, a few federal district courts have adopted similar rules to streamline e-discovery in patent cases, according to legal commentators we spoke with. Courts with e-discovery rules are more efficient and reduce litigation costs for everyone involved, according to a few judges we spoke with. For example, under the model order, discovery begins with an exchange of core documentation concerning the patent, allegedly infringing product, prior art, and finances before an exchange of e-mails.[71] However, at least one federal district court that handles a large volume of patent infringement cases declined to adopt specific discovery rules, noting that it would be a mistake for the courts to bind themselves to specific rules given differences among cases.

In general, several stakeholders we spoke with agreed that it is too early to tell what impact e-discovery rules will ultimately have on patent litigation.

Evolving Case Law

Several federal judges and other stakeholders we interviewed said that the judiciary had the ability to address some of the aforementioned factors contributing to patent infringement litigation in court decisions and that certain recent court decisions have helped to address some of these problems. In the future, they said the courts would continue working to address them. Specifically, these stakeholders said that recent court decisions about what constitutes patentable subject matter and what satisfies the obviousness and definiteness requirements would help combat overly broad and ill-defined patent claims once the decisions were incorporated into PTO's patent examination guidelines and implemented by examiners. For example, two stakeholders said that a 2012 Supreme Court decision restricted what constitutes a patentable process and a 2007 Supreme Court decision made it easier to combine separate pieces of prior art to prove a patented invention was obvious and, thus, not eligible for a patent.[72] In addition, as we discussed earlier, some stakeholders also identified several decisions that have changed damages law. Despite these cases, some stakeholders said that the judicial system has contributed to recent problems in patent infringement litigation and is either unable or unwilling to rectify them. For example, they noted that judges should be more willing to award attorney fees in exceptional cases to the winning party in order to address the growing number of patent infringement lawsuits.

PTO HAS TAKEN ACTIONS TO IMPROVE PATENT QUALITY AND IMPLEMENT NEW AIA PROCEEDINGS THAT MAY AFFECT PATENT LITIGATION IN THE FUTURE

PTO has taken several recent actions that are likely to affect patent quality and patent litigation in the future, including agency initiatives and changes required by AIA. For example, PTO is implementing initiatives to continue to improve the quality of software-related patents and to improve patent examination searches. The agency is also implementing the new administrative trial proceedings created by AIA.

PTO Is Implementing Initiatives to Address Problems with Unclear Patents and Patent Search

We identified four initiatives that PTO is currently undertaking to address patent quality and patent search:

- In February 2011, PTO issued supplemental guidelines to assist in the examination of claims in applications for compliance with the definiteness requirement and began training patent examiners in how to implement them. PTO often updates its patent examination manual or issues guidance to patent examiners in response to changes in case law and these recently issued supplemental guidelines try to ensure that all technologies receive consistent examination practices, according to PTO officials. These guidelines specifically address examination of claims with functional language—which recites a feature by what it does rather than by what it is—in computer-implemented claims. For example, the guidelines state that examiners should determine whether the application discloses the computer and algorithm that perform the claimed function in sufficient detail that someone of ordinary skill in the art would know how to program the computer to perform the claimed function, and that the inventor was in possession of the claimed invention. The guidelines further noted that the algorithm may be expressed in any understandable terms and generally do not call for the application to contain actual computer code. The supplemental guidelines attempt to make examination of applications consistent and resulting patents clearer across all technologies, according to PTO officials. In May 2012, an internal PTO review showed that PTO examiners were making 4 to 6 percent more rejections in patent applications across all technology areas based on claims not being clear and definite than before the supplemental guidelines were issued. PTO officials said patent requirements are established by statute and case law, which constrains the agency's ability to alter requirements for patent applications. PTO officials told us that only court decisions or statutory changes could change the law to require more detail for algorithms in software-related patents, for example.

- In November 2011, in response to the recommendations from FTC's 2011 report, PTO acknowledged that more uniform terminology would help to improve the quality of software-related patents and began working to establish a partnership with the software industry to address this issue.[73] In January 2013, PTO announced the Software Partnership—a cooperative effort with the software industry to explore ways to enhance the quality of software-related patents, including through the use of more uniform terminology. Subsequently, roundtable meetings with the software industry were held in New York City, New York and Stanford, California in February 2013.[74]
- Additionally, PTO is working to implement a new patent classification system, called the Cooperative Patent Classification (CPC), which launched in January 2013. Since October 2010, PTO and the European Patent Office (EPO) have been collaborating to develop the CPC, which is a joint patent classification system that is intended to allow companies and patent applicants to conduct more effective searches for patents that might be related to technologies they are developing or planning to use, which may reduce some of the infringement that contributes to current patent litigation, according to documents we reviewed. CPC is also designed to allow for more frequent updating of patent classes, so that similar technologies can be grouped together, which should improve PTO's ability to adapt to

emerging technologies quickly, according to agency documents we reviewed.

- PTO is also working to provide greater transparency on patent ownership by and issued a notice in November 2012 to obtain public input on regulations it was considering to require patent ownership information to be verified and updated at certain times during patent examination and after patent issuance. Specifically, PTO is currently considering issuing regulations that would require ownership information to be verified and updated periodically. In addition, PTO held a round table discussion in January 2013 to obtain public input on how the agency could collect and provide such information to the public. Since holding the round table, PTO officials have been considering the input and written comments and have been discussing what the appropriate next steps should be. PTO officials noted, however, that the agency does not have substantive rulemaking authority so its ability to issue regulations requiring updates to the ownership information is limited without additional statutory authority.

In addition to these initiatives, in 2009 PTO began a patent quality initiative to measure and improve overall patent quality. After public comment and round table discussions, PTO released a set of new metrics in 2010 for fiscal year 2011 to assess patent quality throughout examination, rather than solely at the end of examination, which had been its practice. These metrics include, among other things, looking at the timeliness of patent examiners' decisions and whether examiners reject claims for the right reasons.[75] However, the agency generally does not review patent examination quality after patents are issued. PTO officials said that they did not examine the nature of patent infringement litigation and issues in dispute or review trends in such litigation. In 2003, the National Academies reported that litigation rates could be a useful measure of post-grant patent quality, presuming that litigated patents define less clear boundaries.[76] One study in the National Academies' report examined the link between examiner inputs, including the number of hours used to examine a patent, and the likelihood that the patent would end up in a lawsuit. The study found a statistically significant result, implying that patents that examiners spend more time examining are less likely to be involved in litigation, suggesting the patents are of higher quality.[77]

In February 2013, PTO officials said that they generally try to adapt to developments in patent law and industry to improve patent quality, as they are doing with their new examination guidelines and their software-related patent partnership. However, the agency does not currently use information on patent litigation in initiating actions like these. However, some staff in PTO's Office of the Chief Economist have suggested that analyzing relationships between the types of patents involved in infringement litigation and internal patent examination data, including the timeline between the filing and grant of a patent and changes in the wording of claims could potentially benefit patent quality and examination by identifying meaningful patterns in the examination of patents that end up in court.[78] As noted, our analysis of litigation data showed that about two-thirds of patent infringement defendants were sued for infringing software-related patents. Given the extent of litigation associated with these patents, examining trends in patent infringement litigation, including the types of patents and issues in dispute, and linking this information to internal data on patent examination, could provide PTO with information to improve patent quality and the examination process. However, PTO staff economists said that looking at litigation trends was not without challenges. For example, they said that less than 2 percent of patents end up in court. In addition, patents usually end up in court long after they are issued, and examination procedures may have long since changed. PTO officials and a few stakeholders also noted that there are other factors that affect litigation trends, including economic conditions and inherent differences among industries.

PTO Is Implementing AIA Post-Grant Review Proceedings

PTO is in the process of implementing the three new post-grant review proceedings created by AIA and described previously—inter partes review, post-grant review, and a transitional program for covered business method patents. These proceedings replace or supplement certain reexamination proceedings conducted by patent examiners with proceedings conducted by judges at PTO's Patent Trial and Appeal Board—making them more closely resemble a patent trial, according to PTO officials we spoke with.[79] According to some stakeholders, these proceedings are most likely to be used by companies that have been sued for patent infringement. PTO issued rules to govern the proceedings in August 2012 and hired additional judges soon after AIA was passed. Specifically, PTO's Patent Trial and Appeal Board plans to move from 100 to more than 200 judges on staff by the end of 2013, according to PTO documents we reviewed. In addition, judges at the Board have been coordinating with stakeholders in the patent community to create a streamlined discovery process for these proceedings. The judges told us they want to avoid the massive discovery requests that add time and expense to patent trials in federal district courts, but said that details of how they handle discovery will continue to evolve over time.[80] The director of PTO is required to study and report to Congress on how these and other provisions of AIA are being implemented no later than September 16, 2015. According to agency officials, PTO plans to meet this 2015 deadline, and the agency has started collecting and publishing information on the number of times the new AIA proceedings are being used and the timing of implementation. As of July 31st, 2013, there had been 395 requests for inter partes reviews and 39 requests for review under the transitional program for covered business method patents, according to data from PTO.

Given that challenging a patent's validity is a common defense for those who have been accused of patent infringement, PTO officials we spoke with mentioned that the new post-grant proceedings may have the potential to reduce future patent litigation because they offer a less costly and faster alternative to settling patent disputes in federal district courts. For example, a representative of a number of technology companies we spoke with estimated that the legal and filing fees for using one of these proceedings would be between \$166,000 and \$390,000, depending on the proceeding, for an average patent, while legal fees could be upwards of \$5 million for an infringement case filed in district court. According to the Board judges we spoke with, these proceedings will be faster than most litigation in district court—which takes 2-1/2 years for the patent infringement trial to begin on average, according to a recent study— because the regulations stipulate that they normally are to conclude within 1 year of being initiated.[81] However, representatives from some operating companies said that they will not use these post-grant review proceedings if federal judges do not consistently delay the costly trials until the Board completes its review.[82] In the past, federal judges have not always been willing to suspend a lawsuit while waiting for PTO to conduct reexaminations, but may be more likely to do this given the regulatory 1 year timeline, according to these representatives. Judges at the Board told us they are working to ensure that the post-grant review proceedings are completed within 1 year. In addition, some operating company representatives we spoke with said that post-grant review in particular would be of limited utility since they only have 9 months after a patent is issued to file a request for review, but usually only learn about a patent's existence after being sued or receiving a demand letter, which is often many years after the patent has been issued.

Finally, representatives from some operating companies said they are not planning to use post-grant review to challenge a patent's validity because an adverse final decision by the Board generally prohibits them from challenging the patent's validity again in court.[83] These representatives said that they consider this to be a major flaw in post-grant review. However, a few other stakeholders said that patent owners need certainty in their patent's validity and that the same party should not be able to challenge the patent's validity through post-grant review and in the federal courts. Judges at the Board said they were aware of this issue and would continue to monitor it.

CONCLUSION

Public discussion surrounding patent infringement litigation often focuses on the increasing role of NPEs. However, our analysis indicates that regardless of the type of litigant, lawsuits involving software-related patents accounted for about 89 percent of the increase in defendants between 2007 and 2011, and most of the suits brought by PMEs involved software-related patents. This suggests that the focus on the identity of the litigant—rather than the type of patent—may be misplaced. PTO's recent efforts to work with the software industry to more uniformly define software terminology and make it easier to identify relevant patents and patent owners may strengthen the U.S. patent system. Further, PTO has available internal data on the patent examination process that could be linked to litigation data, and a 2003 National Academies study reported that using these types of data together could provide useful insights into patent quality. Examining the types of patents and issues in dispute represents a potentially valuable opportunity to improve the quality of issued patents and the patent examination process and to further strengthen the U.S. patent system.

RECOMMENDATION FOR EXECUTIVE ACTION

We are recommending that the Secretary of Commerce direct the Director of PTO to consider examining trends in patent infringement litigation, including the types of patents and issues in dispute, and to consider linking this information to internal data on patent examination to improve the quality of issued patents and the patent examination process.

AGENCY COMMENTS AND OUR EVALUATION

We provided a copy of our draft report to PTO for review and comment. PTO concurred with key findings and our recommendation. PTO also provided additional clarifying comments, which we incorporated as appropriate.

Frank Rusco
Director,
Natural Resources and Environment

APPENDIX I: OBJECTIVES, SCOPE, AND METHODOLOGY

Section 34 of the Leahy-Smith America Invents Act (AIA)[1] mandated that GAO conduct a study on the consequences of patent litigation by nonpracticing entities (NPE).[2] Our objectives were to determine: (1) what is known about the volume and characteristics of recent patent litigation activity; (2) what is known about the key factors that contribute to recent patent litigation trends; (3) what developments in the judicial system may affect patent litigation; and (4) what actions, if any, has the Patent and Trademark Office (PTO) recently taken that may affect patent litigation in the future.

To address all four of these objectives, we reviewed relevant laws, including AIA, and interviewed officials from PTO, the Federal Trade Commission (FTC), and the International Trade Commission (ITC). We also interviewed the following 44 stakeholders knowledgeable about patent litigation:

- Ten representatives from operating companies and industry groups from an array of industries, including software, computer hardware, retailers, and pharmaceuticals, who had been regularly sued in recent years for patent infringement. We identified these operating companies and industry groups by using patent infringement litigation data from 2005 through 2011, which we discuss below.
- Eight representatives of patent monetization entities (PME) and research firms that had regularly sued others in the past 10 years. We identified these PMEs by using patent infringement litigation data

from 2005 through 2011, and they represent a range of business models.

- Fourteen legal commentators, economists, and consultants that had conducted research closely related to our objectives. We identified them through our review of academic literature on patent litigation and the patent system and they each had done work related to the issues we were asked to study.
- We also interviewed one representative each from two large research universities that license patents, two patent brokers who help others buy and sell patents, four venture capitalists, and four individual inventors at software start-up companies and at a small inventor advocacy group. We identified these universities, brokers, venture capitalists, and start-up companies based on our review of academic literature on patent litigation and the patent system and they were knowledgeable of the issues we were asked to study.

Because stakeholders varied in their expertise with various topics, not every stakeholder provided an opinion on every topic.

We were not able to find reliable data on patent assertion outside of the court system—which often consist of patent owners sending letters demanding licensing fees to potential infringers before filing suit and, as a result, our data analysis focuses on patent infringement litigation rather than patent assertion more broadly.

In addition to steps we took to address all four objectives, to describe what is known about the volume and characteristics of recent patent infringement litigation activity, we reported data from the American Intellectual Property Law Association's (AIPLA) 2011 survey on the costs of patent litigation.[3] We also reviewed academic literature on patent litigation and the patent system in general and assessed the methodology of the studies we reported on for soundness. To assess the reliability of data from PTO, AIPLA, and RPX, we conducted interviews and reviewed relevant methodology documentation. We found that these data were sufficiently reliable for purposes of this report.

We also analyzed patent infringement litigation data that we purchased from Lex Machina, a firm that collects and analyzes data on patent litigation. 4 Lex Machina maintains a database created from public electronic court filings for all patent infringement lawsuits filed in U.S. federal district courts, beginning in the year 2000. From this database, Lex Machina selected a random, generalizable sample of 100 lawsuits per year from 2007 through 2011, identifying for each lawsuit the patent(s) being litigated, the court hearing the lawsuit, and the lawsuit's outcome. The sample size allowed us to draw conclusions about patent infringement lawsuits filed in each of these years, with a margin of error of no more than plus or minus 5 percentage points over all these years and no more than plus or minus 10 percentage points for any particular year.[5] Each sample element was subsequently weighted in the analysis to account statistically for all the members of the population, including those that were not selected. To assess the reliability of data from Lex Machina, we met with Lex Machina staff, examined documentation, and tested and reviewed the sample data for completeness and accuracy. We found the data to be sufficiently reliable for our purposes. We also obtained patent infringement data from RPX, another firm that collects data on patent infringement lawsuits, in an effort to verify Lex Machina's litigant categorizations.[6]

For the plaintiffs named in each lawsuit, Lex Machina also provided an entity description, classifying the plaintiffs as follows:

- an operating company, or a likely operating company;
- an entity related to an operating company;
- a PME or a likely PME;
- a university;
- an individual or trust; or
- a research firm.[7]

To classify a plaintiff, Lex Machina placed more weight on statements made by the plaintiff in official documents, such as its own court and Securities and Exchange Commission (SEC) filings, and statements appearing on a website maintained by the plaintiff. From these sources Lex Machina obtained information that would help it to determine the entity type, such as statements indicating that the plaintiff made or sold a product or offered services, or evidence that the plaintiff shared corporate ownership with such an entity. Some statements, such as a statement describing a plaintiff's business as focused on patent licensing, indicated that the plaintiff was in the business of patent monetization.

In some cases, the plaintiffs' own statements did not lead to a definitive classification, and Lex Machina consulted other sources to obtain relevant information about the plaintiff's business activities. To the extent possible, Lex Machina relied on additional sources they characterized as "verified and objective," such as corporate databases and articles of incorporation. To the extent that Lex Machina lacked official information, it used professional judgment to classify plaintiffs based on information from other sources, such as news articles, blogs, court filings by opposing parties, and Lex Machina's own database. Examples of information from these other sources that were used to classify plaintiffs include: (1) the number of patent infringement lawsuits filed by the plaintiff; (2) whether the plaintiff had been sued for patent infringement in other lawsuits or counterclaims; (3) whether the plaintiff had been subject to a declaratory judgment lawsuit for noninfringement or invalidity; (4) the number of defendants sued in each lawsuit and in total; (5) statements about the plaintiff in news articles, blogs, or in court filings by an opposing party, such as a statement characterizing the plaintiff as a patent monetization entity; (6) whether the plaintiff was represented by an attorney known to represent patent monetization entities; and (7) evidence of linkages between the plaintiff and a patent monetization entity, such as a common principal or a street address shared in common. In some cases, the difficulty of finding information about the plaintiffs, their business activities, and their corporate relationships led Lex Machina to choose no entity classification at all. Such plaintiffs were classified as "Insufficient Evidence."

We reviewed Lex Machina's classifications of first named plaintiffs to ensure their reliability and consistency. We found 29 cases where we differed with Lex Machina's original classification. They adjusted their classifications in all but five of the cases. We agreed on different labels in these cases mainly based on our preference for the "entity related to operating company" label. Table 1 presents the evidence or information we used to review Lex Machina's classifications of first named plaintiffs.

Table 1. Evidence or Information Used to Review Lex Machina's Plaintiff Classifications

Plaintiff classification	Evidence or information used to review plaintiff classification
Operating company	The plaintiff's website, court filings, or SEC filings indicated that the plaintiff made a product,sold a product, or offered a service at the time the lawsuit was filed.
Likely operating company	In the absence of information from the plaintiff's website, court filings, or SEC filings, other sources indicated that the plaintiff made a product, sold a product, or offered a service at the time the lawsuit was filed.

Table 1. (Continued)

Plaintiff classification	Evidence or information used to review plaintiff classification
Entity related to an operating company	• There was no available evidence that the plaintiff made a product, sold a product, or offered a service but there was evidence that its parent corporation, subsidiary, or otherrelated entity made a product, sold a product, or offered a service; or • The plaintiff was a holding company that did not itself make a product, sell a product, oroffer a service but owned the stock of a company that did; or • The plaintiff was a parent company to at least one subsidiary that made a product, sold a product, or offered a service; or • The plaintiff's business consisted of patent monetization but its parent, subsidiary, or other company with shared corporate ownership made a product, sold a product, or offered a service. • This category did not include plaintiffs if the only known association between it and the operating company was a licensing agreement.
Patent monetization entity	• The plaintiff indicated that it, or a corporate parent, was in the patent monetization business, by stating that its or its parent's purpose was to license patents or generate revenue from licensing patents; or • There was no available evidence of the plaintiff making, selling, or offering a product orservice, and the plaintiff had never been sued for patent infringement, and the plaintiff had filed 8 or more patent infringement lawsuits;aor • There was no available evidence of the plaintiff making, selling, or offering a product orservice, and the plaintiff had never been sued for patent infringement, and the plaintiff had sued a large number of defendants (around 20 or more, which put them in the top 5 percent of cases); or • There was no available evidence of the plaintiff making, selling, or offering a product orservice, the plaintiff had never been sued for patent infringement, and there were at least 4 pieces of different types of other evidence supporting classification as a patent monetization entity.
Likely patent monetization entity	• There was no available evidence of the plaintiff making, selling, or offering a product orservice, the plaintiff had not been sued for patent infringement, and there were 2 or 3 pieces of different types of other evidence supporting the plaintiff's classification as a patent monetization entity.

Plaintiff classification	Evidence or information used to review plaintiff classification
University	The plaintiff is an institution of higher education.
Individual/trust	The plaintiff identified itself in court documents as an individual or a trust.
Research firm	The plaintiff's business focus or purpose was on research and development, even if it also produced or marketed a product or service.
Insufficient evidence	• There was conflicting information about whether, at the time the lawsuit was filed, the plaintiff made or sold a product, offered a service, or was in the business of patent monetization; or • There was no more than 1 piece of other evidence supporting the plaintiff's classification as a patent monetization entity.

Source: GAO.

[a]The 106 most-asserted patents from January 2000 through February 2009 were litigated eight times or more. See John R. Allison et al., *Patent Quality and Settlement Among Repeat Patent Litigants,* 99 Geo. L. J. (2011).

In addition to steps we took to address all four objectives, to describe what is known about the key factors that contribute to recent patent litigation trends, we reviewed academic literature on the patent and judicial system and the benefits and costs of patent assertion, including economic and legal studies.

In addition to steps we took to address all four objectives, to describe developments in the judicial system that may affect patent litigation, we interviewed officials and judges from the U.S. District Courts for the District of Delaware and for the Eastern District of Texas. We selected these district courts because they had high levels of patent infringement lawsuits according to Lex Machina data. We also interviewed judges with the U.S. Court of Appeals for the Federal Circuit in Washington, D.C., which hears appeals of patent cases decided in federal district courts. We also interviewed officials from the Administrative Office of the U.S. Courts (AOUSC), and the Federal Judicial Center—organizations that provide broad administrative, legal, and technological services and support to the judicial branch. We also reviewed documents and data from the courts, as well as economic and legal studies.

In addition to steps we took to address all four objectives, to describe what actions, if any, PTO has recently taken that may affect patent litigation in the future, we conducted interviews with officials from PTO and reviewed documents and data from this agency, as well as economic and legal studies.

End Notes

[1] A patent is an exclusive right granted for a fixed period of time to someone who invents or discovers (1) a new and useful process, machine, manufacture, or composition of matter or (2) any new and useful improvement of such items.

[2] In this report, when we use the term "patent owner," it includes the real party in interest when that party is not the patent owner. A real party in interest may be, among other things, an entity that has a legal right to enforce the patent, such as a parent entity or exclusive licensee.

[3] Anyone who makes, sells, offers to sell, uses, or imports the patented invention during the term of the patent without the patent owner's permission infringes the patent. Patent infringement is a strict liability offense—the alleged infringer's intent to copy or act of copying the patented invention are not relevant to the outcome of an infringement lawsuit—so an individual who independently develops an invention that falls within the scope of a patent may infringe the patent. A patent owner can grant permission to use a patented invention by licensing others to use, make, sell, or import the patented invention. Patent owners can also transfer title to their patents by assigning their patent rights to others.

[4] For more information on the dispute between Howe and Singer, see: Adam Mossof, *The Rise and Fall of the First American Patent Thicket: The Sewing Machine War of the 1850s*, 53 Ariz. L. Rev.165 (2011). See also Ryan L. Lampe & Petra Moser, *Do Patent Pools Encourage Innovation? Evidence from the 19th-Century Sewing Machine Industry,* NBER Working Paper No. 15061 (June 2009).

[5] The Constitution grants to Congress the power "[t]o promote the Progress of Science and useful Arts, by securing for limited Times to Authors and Inventors the exclusive Right to their respective Writings and Discoveries." U.S. Const., art. I, § 8, cl. 8.

[6] The Federal Trade Commission uses the related term "patent assertion entities" to focus on entities whose business model solely focuses on asserting typically purchased patents. As such, the PME term also encompasses entities that might use third-party NPEs to assert patents for them.

[7] FTC's mission includes prevention of and enforcement against anticompetitive, unfair, or deceptive business practices including, potentially, patent assertion activities.

[8] This is not unique to patent infringement litigation. As discussed later in the report, in civil lawsuits, the parties must exchange certain information relevant to the litigation, a process known as discovery. Discovery costs in complex litigation, including patent infringement litigation, can run into the millions of dollars. Because NPEs do not make products, they generally have less information to disclose and thus have lower discovery costs. They also cannot be countersued for patent infringement. This asymmetry in litigation costs (which exists in other types of complex litigation) can give NPEs leverage in seeking financial compensation from operating companies.

[9] Pub. L. No.112-29 § 34 (2011).

[10] As noted in a September 7, 2011, letter from the Comptroller General to the chairs and ranking members of the congressional committees with jurisdiction over patents, the bill being considered at that time would have required a GAO study involving several questions for which reliable data were not available or could not be obtained. The bill was enacted without change, but the Chair of the Senate Judiciary Committee, responding to these concerns, stated that GAO should note data and methodology limitations in its report prepared in response to the mandate. 157 CONG. REC. S 5402, S5441 (daily ed. Sept. 8, 2011) (statement of Sen. Leahy). Consequently, we developed report objectives consistent

with these limitations, and we have noted specific data limitations in appendix I and throughout this report, as appropriate.

[11] We identified some of these stakeholders from patent infringement litigation data from 2000 through 2011 that we reviewed. Representatives of companies and PMEs we talked with had regularly been sued or had regularly sued others over the past decade. Other stakeholders we identified through our review of academic literature on patent litigation and the patent system and were knowledgeable in the issues we were asked to study. Because stakeholders varied in their expertise with various topics, not every stakeholder provided an opinion on every topic.

[12] This sample allowed us to draw conclusions about the broader population of patent infringement lawsuits for each of these years and is therefore generalizable to all patent infringement lawsuits filed in federal district court from 2007 to 2011. However, as noted, estimates from the Lex Machina sample are subject to a 5 percent margin of error. This means that an estimate of 50 percent, for example, based on all years of data, would have a 95 percent confidence interval of between 45 percent and 55 percent. The margin of error is 10 percent when looking at individual years, which means that an estimate of 50 percent, for example, looking at an individual year, would have a 95 percent confidence interval of between 40 percent and 60 percent. Because Lex Machina followed a probability procedure based on random selections, the sample is only one of a large number of samples that might have been drawn. Since each sample could have provided different estimates, we express our confidence in the precision of our particular sample's results as a 95 percent confidence interval. This is the interval that would contain the actual population value for 95 percent of the samples that could have been drawn. Unless otherwise noted, the margin of error associated with the confidence intervals of our survey estimates is no more than plus or minus 10 percentage points at the 95 percent level of confidence.

[13] Definitions of these categories are discussed below and detailed in appendix I. We found that it was difficult to reliably identify the type of NPEs through analysis of data from court records because, among other things, firms do not identify themselves as such in these records. Lex Machina did not include patent owners that primarily seek to develop and transfer technology, such as universities and research firms, as PMEs. See appendix I for more detail on Lex Machina's categorizations and our review of them.

[14] RPX also purchases patents itself, to prevent them from being asserted against its members. RPX provided us with summary data on the number of patent infringement lawsuits filed in federal district court since January 2005. RPX's data identified NPEs and other types of plaintiffs in these lawsuits by using a variety of factors, such as whether there was evidence that an entity sells or develops products. RPX representatives said that they used professional judgment to some extent in making these determinations. We were not able to fully assess the reliability of the judgments RPX used in making these classifications.

[15] AIPLA is a national, voluntary bar association constituted primarily of patent lawyers in private and corporate practice, in government service, and in the academic community. See AIPLA, *Report of the Economic Survey 2011* (Arlington, Va.: July 2011). AIPLA surveyed its members during 2011 and asked them to estimate legal costs for typical patent infringement cases. AIPLA's findings are based on an 18 percent response rate.

[16] PTO's data show that the current inventory of new applications that have not yet received an initial decision was around 600,000 applications. PTO refers to these initial decisions as a "first action on the merits."

[17] To be patentable subject matter, the invention must be a (1) process; (2) machine; (3) manufacture; (4) composition of matter; or (5) improvement of a process, machine,

manufacture, or composition of matter. To be nonobvious, the claimed invention's improvements to the prior art must be more than the predictable use of prior art elements according to their established functions. Specifically, at the time of the invention, the differences between the scope and content of claimed invention and the prior art cannot render the claimed invention as a whole obvious to a person having ordinary skill in the art.

[18] During the patent examination, the applicant and the examiner communicate about the application, including aspects that might be deficient. For example, the examiner may inform the applicant that one of the claims is not novel because of prior art, and the applicant might revise the claim to distinguish it from the prior art the examiner found.

[19] Inter partes is Latin for "between the parties."

[20] A covered business method patent is a patent that claims a method or corresponding apparatus for performing data processing or other operations used in the practice, administration, or management of a financial product or service but does not include patents for technological inventions. This transitional program is subject to a sunset provision that will repeal the program on September 16, 2020.

[21] In patent infringement lawsuits, the accused infringer often challenges the patent's validity as an "affirmative defense," meaning that even if the infringement allegations are true, the would-be infringer is not liable because the patent is invalid. A party accused of infringement can also file a lawsuit to obtain a court decision on whether they are infringing or whether the patent is valid, which is known as a declaratory judgment action.

[22] As noted, asymmetrical discovery demands, burdens, and costs are not unique to NPE patent infringement litigation. For example, parties in class actions and antitrust litigation typically face the same asymmetry. *See, e.g.,Thorogood v. Sears, Roebuck & Co.*, 624 F.3d 842, 849-50 (7th Cir. 2010) (class action discovery); *Bell Atlantic Corp. v. Twombly*, 550 U.S. 544, 558-559 (2007) (antitrust discovery).

[23] This hearing is often referred to as a *Markman* hearing after the Supreme Court's decision in *Markman v. Westview Instruments, Inc.*, 517 U.S. 370 (1996).

[24] The judge has the authority to increase damages up to three times the amount initially awarded, called treble damages, for cases of willful infringement. In "exceptional cases" the court is authorized to award the prevailing party reasonable attorney fees.

[25] In order for ITC to have jurisdiction to hear the complaint, there must be an industry in the United States in existence or in the process of being established that relates to the articles protected by the patent concerned. An industry in the United States is considered to exist if there is in the country, with respect to the articles protected by the patent: (1) a significant investment in plant and equipment; (2) significant employment of labor or capital; or (3) substantial investment in its exploitation, including engineering, research and development, or licensing. This is known as the domestic industry requirement.

[26] Prior to this 2006 Supreme Court case, the Federal Circuit's general rule was for district court judges to issue injunctions in patent cases once validity and infringement had been determined except in unusual cases under exceptional circumstances and in rare instances to protect public interest. In the 2006 *eBay* decision, the Supreme Court ruled that district courts should not assume an injunction was automatically needed in patent infringement cases and instead should use the same test used in other cases to determine whether to award the plaintiff an injunction. *eBay, Inc. v. MercExchange, L.L.C.*, 547 U.S. 388 (2006). According to several legal commentators we spoke with, this decision has generally made it more difficult for NPEs to obtain injunctions in the courts and has led them to pursue exclusion orders at ITC—although there may have been other reasons for the increase in filings, including the relative speed of proceedings at ITC.

[27] Although PTO does not have a specific "software-related" patent class, we combined a number of entire patent classes that PTO economists have said are most likely to include patents with software-related claims, and this includes business method patents. For the list of these classes, see S. Graham, , and S. Vishnubhakat, *Of Smart Phone Wars and Software Patents,* Journal of Economic Perspectives, v. 27 no.1 (2013), pp. 67-86.

[28] *Gottschalk v. Benson*, 409 U.S. 63 (1972) (finding a mathematical formula that had no substantial practical application except in connection with a digital computer was not patentable because it is like a law of nature, which cannot be the subject of a patent); *Parker v. Flook,* 437 U.S. 584 (1978) (finding a method for updating alarm limits through computerized calculations was not patentable because the alarm limit is a number and the patent application was for a formula to compute it).

[29] *Diamond v. Diehr*, 450 U.S. 175 (1981) (finding a patent claim containing a mathematical formula that implements or applies that formula in a structure or process, which, when considered as a whole, is performing a function which the patent laws were designed to protect, to be patentable).

[30] *State Street Bank & Trust Co. v. Signature Financial Group, Inc.*, 149 F.3d 1368 (Fed. Cir. 1998).

[31] *Bilski v. Kappos*, __ U.S. __, 130 S. Ct. 3218 (2010).

[32] Pub. L. No. 112-29, § 19(d)(1) (2011). Specifically, accused infringers may be joined, or have their actions consolidated for trial, only if (1) questions of fact common to all will arise in the action; and (2) any right to relief is asserted against the parties jointly and severally or with respect to or arising out of the same transaction, occurrence, or series of transactions or occurrences relating to the same accused product or process. AIA Section 19(d)(1), however, allows an accused infringer to waive these restrictions.

[33] H. Rep. No. 112-98, at 54 (2011).

[34] The evidence that Lex Machina used to classify an entity as an operating company and that we then used to review Lex Machina's classifications is described in appendix I. Related entities included subsidiaries of operating companies—see appendix I for more information. We did not verify whether the companies practiced the patents at issue in the lawsuit.

[35] The evidence that Lex Machina used to classify an entity as a PME or likely PME and that we then used to review Lex Machina's classifications is described in appendix I. Another paper using data from Lex Machina presented different proportions of patent monetizing plaintiffs, and these differences may be due to differences in methodology. For example, this study included other plaintiff groups as patent monetizers, including individuals and trusts. See Robin Feldman, Tom Ewing, and Sara Jeruss, *The AIA 500 Expanded: The Effects of Patent Monetization Entities*, UCLA Journal of Law & Technology (forthcoming).

[36] Our analysis of litigation data from RPX showed similar results. Specifically, RPX's classification of all infringement suits from 2007 to 2011 shows that operating companies brought 69 percent of lawsuits, and firms that RPX classified as patent assertion entities (PAE) brought 25 percent. RPX's PAE category excluded universities and individual inventors acting as NPEs, making it similar to Lex Machina's PME category. Individual inventors brought about 6 percent of the lawsuits, and universities brought less than 1 percent. Operating companies litigating patents that do not relate to the technology of their primary business sector—classified as noncompeting entities by RPX—brought less than 1 percent of all lawsuits. Additionally, lawsuits filed by PAEs increased by about a third from 2007 to 2010 and, in 2011, doubled over the previous year, and lawsuits brought by

operating companies decreased by about 6 percent from 2007 to 2010 and, in 2011, increased by about 3 percent over the previous year.

[37] Our analysis of RPX data showed similar results. Specifically, RPX's data show that from 2007 to 2011 operating companies sued about 51 percent of all defendants and that firms RPX classified as PAEs sued about 42 percent of all defendants. Additionally, defendants sued by these firms almost tripled, while defendants sued by operating companies decreased by about 20 percent. RPX identified more than 200 NPEs that sued more than 20 defendants since 2005, suing 12 defendants per suit on average.

[38] Often both the NPE and operating company appear as plaintiffs in the same suit.

[39] For example, some operating companies owned patent monetization subsidiaries that shared their name, and others were linked to outside monetization entities with different names.

[40] RPX collects data on these types of lawsuits and, according to RPX's data, these lawsuits accounted for less than 1 percent of lawsuits in RPX's database from 2007 to 2011.

[41] These research firms told us that they file patent infringement lawsuits if their patented technologies are used without or in violation of licensing agreements. Economic literature suggests that this division of labor is valuable because companies that specialize in R&D can often innovate more nimbly and create new, innovative technologies that other companies can then incorporate into products they manufacture.

[42] We estimate that, from 2007 to 2011, operating companies and related entities brought 3,037 lawsuits involving software-related patents, while PMEs and likely PMEs brought 2,093 such lawsuits.

[43] Defendants can request that the district court transfer the case to another court in certain circumstances. If the district court denies the defendant's request, the defendant can ask the Federal Circuit to order the district court to transfer the lawsuit. For example, in 2008, the Federal Circuit ordered the Eastern District of Texas to transfer a case because none of the parties had an office and no witnesses or evidence were located in the district. *In re: TS Tech*, 551 F.3d 1315 (Fed. Cir. 2008). According to one legal commentator, the precedent in this case and other Federal Circuit transfers likely resulted in fewer lawsuits being filed in the Eastern District of Texas.

[44] PricewaterhouseCoopers, *2012 Patent Litigation Study: Litigation continues to rise amid growing awareness of patent value. www.pwc.com/us/en/forensicservices/publications/ 2012-patent-litigation-study.jhtml.* Pricewaterhouse looked at decisions identified through the WestLaw database.

[45] We estimate that 46 percent of all patent infringement lawsuits filed between 2007 and 2011 ended or likely ended in a settlement within 1 year of being filed.

[46] Of the lawsuits that were not ongoing, 86 percent of PME suits and 87 percent of operating company suits settled, which was not a statistically significant difference. However, our analysis showed a statistically significant difference between suits involving software-related patents, of which 82 percent settled compared with 89 percent of suits that did not involve software-related patents.

[47] For the 18 percent of those lawyers that responded, AIPLA reports that the median legal cost for one patent infringement lawsuit was $650,000 when less than $1 million was at risk for damages; $2.5 million when between $1 million and $25 million was at risk for damages; and $5 million when more than $25 million was at risk for damages. These costs include legal fees and exclude damage awards.

[48] PricewaterhouseCoopers, *2012 Patent Litigation Study: Litigation continues to rise amid growing awareness of patent value.* Pricewaterhouse looked at decisions identified through the WestLaw database.

[49] James E. Bessen and Michael J. Meurer, *The Direct Costs from NPE Disputes,* Law and Economics Research Paper No. 12-34 (Boston University School of Law: June 28, 2012).

[50] Colleen Chien, *Startups and Patent Trolls,* Santa Clara Univ. Legal Studies Research Paper No. 09-12 (Sept. 28, 2012).

[51] The author of a 2012 study of a nonrandom, nongeneralizable set of 223 small technology companies noted above found that about two-thirds of the 79 companies that reported that they had received a demand from an NPE were not sued.

[52] *MedImmune Inc. v. Genentech*, 549 U.S. 118 (2007).

[53] In addition, FTC recently reported that reliance on describing what an invention does rather than how it is created contributes to overly broad patents and a 2012 study also supports that view. U.S. Federal Trade Commission, *The Evolving IP Marketplace: Aligning Patent Notice and Remedies with Competition* (Washington, D.C.: March 2011). Mark A. Lemley, *Software Patents and the Return of Functional Claiming* (Stanford University, 2012).

[54] Applications for reissuance of defective patents—those which, through error, are deemed wholly or partly inoperative or invalid by reason of a defective specification or drawing or by reason of the patentee claiming more or less than he or she had a right to claim in the patent—cannot introduce new matter and must be filed within 2 years of the patent issuance if the reissued patent enlarges the scope of the original patent's claims.

[55] PTO recognizes three types of continuations and only one type allows the patent owner to claim subject matter not previously disclosed in the application. One patent owner told us he filed a continuation so that his patent would closely match software that was already being used but that his original application did not anticipate; he later won large settlements from the allegedly infringing companies.

[56] One study we reviewed looked at a sample of 193 complaints from two federal courts districts, the District of Delaware and the Eastern District of Texas, and found that allegations of copying occurred in 21 of the 193 complaints. The authors suggested this means that most accused infringers came up with the idea independently; however, as noted, patent infringement liability does not require proof of copying so patent owners do not need to allege it. Christopher A. Cotropia and Mark A. Lemley, *Copying in Patent Law,* 87 N.C.L. Rev. 1421 (2009).

[57] AIPLA, *Report of the Economic Survey 2011* (Arlington, Va.: July 2011). Entities can seek patent validity and infringement opinions on patents found during a patent search. A patent validity opinion is a legal opinion as to whether a patent is valid and a patent infringement opinion is a legal opinion as to whether a product infringes a patent.

[58] This practice imposes social costs to businesses that are sued in this manner, according to economists. We were not able to identify suits associated with overly broad patents in our analysis of Lex Machina data, or the extent to which overly broad patents were associated with PME lawsuits, and were not able to estimate the overall economic social costs of this type of patent infringement litigation.

[59] As noted earlier, our analysis of litigation data showed that PME suits accounted for about half of the increase in defendants between 2007 and 2011, while suits involving software-related patents accounted for around 89 percent of the increase in defendants.

[60] U.S. Federal Trade Commission, *The Evolving IP Marketplace: Aligning Patent Notice and Remedies with Competition* (Washington, D.C.: March 2011).

[61] *Uniloc USA, Inc., v. Microsoft Corp.*, 632 F.3d 1292 (Fed. Cir. 2011).

[62] Enhanced damages can be awarded for willful infringement but some stakeholders said that recent Federal Circuit decisions made it harder to prove willful infringement and obtain enhanced damages. *Knorr-Bremse Systeme Fuer Nutzfahrzeuge GmbH v. Dana*, 383 F.3d

1337 (Fed. Cir. 2004) (alleged patent infringers who invoke attorney-client privilege or work product protection over advice of counsel regarding whether its activities would be an infringement of a valid patent does not give rise to an adverse interference and failure to obtain legal advice does not give rise to an adverse inference with respect to willful infringement); *In re: Seagate Technology*, 497 F.3d 1360 (Fed. Cir. 2007) (proof of willful infringement permitting enhanced damages requires the patent owner (1) to show by clear and convincing evidence that the infringer acted despite an objectively high likelihood that its actions constituted infringement of a valid patent and (2) demonstrate that this objectively-defined risk was either known to the accused infringer or so obvious that it should have been known but there is no affirmative obligation to obtain opinion of counsel).

[63] *See* Colleen V. Chien, *From Arms Race to Marketplace: The Complex Patent Ecosystem and Its Implications for the Patent System*, 62 Hastings L.J. 297 (2010); Kevin Rivette and David Kline, Rembrandts in the Attic: Unlocking the Hidden Value of Patents, (Boston, Ma.: Harvard Business School Press, 2000).

[64] For example, one PME we spoke with said its business model was to value patents based on legal, financial, and technical factors and then either (1) offer an inventor no upfront payment but a large cut of a settlement with accused infringers, (2) offer the inventor a small payment up front and a medium-sized portion of the settlement, or (3) pay the inventor up front and retain all of the settlement. One legal commentator, however, noted that contingency fee lawsuits are not unique to patent infringement litigation.

[65] Pub. L. No. 111-349 (2011).

[66] To be eligible for designation, district courts had to: (1) be among the 15 district courts in which the largest number of patent cases were filed in 2010; or (2) have adopted, or intend to adopt, local rules for patent cases.

[67] The law requires patent cases to be randomly assigned to judges in districts participating in the pilot program regardless of whether they are designated by the court's chief judge to hear patent cases. Nondesignated judges may decline to accept the case, which would then be randomly assigned to one of the designated judges.

[68] The law also requires AOUSC to keep certain congressional committees informed on a periodic basis while the pilot project is in effect and to submit a report to those committees in 2016 and 2021.

[69] The Advisory Council was established to review, study, and make recommendations on the rules of practice and internal operating procedures of the court. The Chief Judge of the court unveiled the model order at a meeting of the Federal Circuit Bar Association and the Eastern District of Texas Bar Association. The model order is a template, or model, for judges to adopt and issue in patent infringement lawsuits.

[70] AIPLA, *Report of the Economic Survey 2011* (Arlington, Va.: July 2011). The estimated cost of litigation through discovery was $350,000 when the damage stakes are less than $1 million, $1.5 million when the stakes are between $1 million and $25 million, and $3 million when the stakes are over $25 million.

[71] The model order presumptively limits the number of each party's e-mail requests, but the parties may jointly agree to modify the limits or request court modification for good cause.

[72] *Mayo Collaborative Servs. v. Prometheus Labs*, _ U.S. _, 132 S. Ct. 1289 (2012); *KSR International Co. v. Teleflex, Inc.*, 550 U.S. 398 (2007).

[73] U.S. Federal Trade Commission, *The Evolving IP Marketplace: Aligning Patent Notice and Remedies with Competition* (Washington, D.C.: March 2011), (FTC Report Recommends Improvements in Patent System to Promote Innovation and Benefit Consumers).

[74] PTO announced these public meetings by public notice. 78 Fed. Reg. 292 (Jan. 3, 2013).

[75] According to PTO, the new metrics expand the previous procedures for measuring examination quality and will reveal quality issues arising during the examination and identify their sources so that problems may be remediated by training. See the following for more information on PTO's patent quality metrics: http://www.uspto.gov/dashboards /patents/main.dashxml.

[76] National Academy of Sciences, Committee on Intellectual Property Rights in the Knowledge-Based Economy, National Research Council, *Patents in the Knowledge-Based Economy,* Wesley M. Cohen and Stephen A. Merrill, eds. (Washington, D.C., National Academies Press, 2003).

[77] The results suggested that an additional hour of patent examination per application would be associated with a decrease in litigation by as many as 24 to 26 infringement complaints annually. The study, using data from 1996 and 1997, estimated that the savings associated with this reduction in suits would have been greater than the cost of the additional examiner hours.

[78] PTO's Office of the Chief Economist was created in 2010. See also J. Lanjouw. and M. Schankerman, *Characteristics of Patent Litigation: A Window on Competition,* The RAND Journal of Economics, vol. 32, no. 1 (Spring 2001), pp. 129-151.

[79] Prior to AIA's provisions taking effect on September 16, 2012, patent reexamination proceedings consisted of:(1) inter partes reexamination conducted by patent examiners, which AIA replaced with inter partes review, a proceeding conducted by the Patent Trial and Appeal Board and (2) ex-parte reexamination.

[80] In addition, judges at the Board told us they are working with stakeholders in the patent community to balance some stakeholders' need for additional pages in their written submissions with the Board's need to complete the post-grant reviews in a timely way.

[81] PricewaterhouseCoopers, *2012 Patent Litigation Study: Litigation continues to rise amid growing awareness of patent value.*

[82] Under AIA, if an entity submits a request for inter partes or post-grant review and then subsequently files a lawsuit, but not a counterclaim, challenging the validity of a patent claim in federal district court, the lawsuit is automatically suspended until certain events occur. In addition, AIA specifies factors judges are to consider in deciding whether to suspend a lawsuit alleging patent infringement at the request of a party because of a transitional covered business method patent proceeding. Requests for inter partes review and post-grant review cannot be granted if the requester (or real party in interest) previously filed a lawsuit challenging the validity of the patent's claims—but not a counterclaim—before the date on which the request is filed.

[83] Specifically, parties who request post-grant or inter partes reviews that result in a final written decision may not assert in federal court or before ITC that a claim is invalid on any ground the party raised or reasonably could have raised during the review. Similarly, parties who receive a final written decision in a transitional program for covered business method patent proceedings are prohibited from subsequently asserting in federal court or before ITC that the claim is invalid on any ground that was raised during the proceeding.

End Notes for Appendix I

[1] Pub. L. No.112-29 § 34 (2011).

[2] As noted in a September 7, 2011, letter from the Comptroller General to the chairs and ranking members of the congressional committees with jurisdiction over patents, the bill being considered at that time would have required a GAO study involving several questions for which reliable data were not available or which could not be obtained. The bill was enacted without change, but the Chair of the Senate Judiciary Committee, responding to these concerns, stated that GAO should note data and methodology limitations in its report prepared in response to the mandate. 157 CONG. REC. S 5402, S5441 (daily ed. Sept. 8, 2011) (statement of Sen. Leahy). Consequently, we developed report objectives consistent with these limitations, and we have noted specific data limitations throughout this report, as appropriate.

[3] AIPLA is a national, voluntary bar association constituted primarily of lawyers in private and corporate practice, in government service, and in the academic community. See AIPLA, *Report of the Economic Survey 2011* (Arlington, Va.: July 2011). AIPLA surveyed patent lawyers during 2011 and asked them to estimate legal costs for typical patent infringement cases. Their findings are based on an 18 percent response rate.

[4] We selected Lex Machina through a competitive contracting process to provide us patent infringement litigation data, select a sample of cases, and classify the plaintiffs in those cases.

[5] This sample allowed us to draw conclusions about the broader population of patent infringement lawsuits for each of these years and is therefore generalizable to all patent infringement lawsuits filed in federal district court from 2007 to 2011. However, as noted, estimates from the Lex Machina sample are subject to a 5 percent margin of error. This means that an estimate of 50 percent, for example, based on all years of data, would have a 95 percent confidence interval of between 45 percent and 55 percent. The margin of error is 10 percent when looking at individual years, which means that an estimate of 50 percent, for example, looking at an individual year, would have a 95 percent confidence interval of between 40 percent and 60 percent. Because Lex Machina followed a probability procedure based on random selections, the sample is only one of a large number of samples that might have been drawn. Since each sample could have provided different estimates, we express our confidence in the precision of our particular sample's results as a 95 percent confidence interval. This is the interval that would contain the actual population value for 95 percent of the samples that could have been drawn. Unless otherwise noted, the margin of error associated with the confidence intervals of our survey estimates is no more than plus or minus 10 percentage points at the 95 percent level of confidence.

[6] RPX also purchases patents itself, to prevent them from being asserted against its members. RPX provided us with summary data on the number of patent infringement lawsuits filed in federal district court since January 2005. RPX's data identified NPEs and other types of plaintiffs in these lawsuits by using a variety of factors, such as whether there was evidence that an entity sells or develops products. RPX representatives said that they used professional judgment to some extent in making these determinations. We were not able to fully assess the reliability of the judgments RPX used in making these classifications.

[7] Lex Machina originally labeled these "other entity," but we determined that research firm was a better descriptive term for these companies, although in a few cases the entity was government sponsored.

INDEX

D

E

F

G

H

I

J

L

M